# Chemical Principles
# in the Laboratory

THIRD EDITION

# Chemical Principles in the Laboratory

**Robert F. Bryan**
*University of Virginia*

**Robert S. Boikess**
*Rutgers, The State University of New Jersey*

**HARPER & ROW, PUBLISHERS, New York**
Cambridge, Philadelphia, San Francisco,
London, Mexico City, São Paulo, Singapore, Sydney

**Sponsoring Editor:** Heidi Udell
**Project Coordination, Art and Design:** Caliber Design Planning, Inc.
**Cover Design:** Hudson River Studios
**Cover Photo:** Robert S. Boikess
**Compositor:** Typo-graphic Alternatives
**Printer and Binder:** Malloy Lithographing, Inc.

The cover illustration is a computer-generated model of a complex consisting of the enzyme carboxyl proteinase and its substrate, the drug pepstatin. The image shows a molecule of pepstatin that is bound to the active site of the enzyme. This kind of high-resolution computer model is now playing an important role in the development of drugs by enabling the design of compounds that are tailored to fit the active sites of biological molecules. Computer models also have become indispensable tools for the study of complex biological molecules such as proteins, nucleic acids, and carbohydrates.

The complex of pepstatin and proteinase was described by R. Bott, E. Subramanian, and D. Davies; the coordinates were provided by Dr. Joel Sussman and Dr. David Davies. The computer graphics were produced with the assistance of Professor R. Levy, Professor W. Olson, and Dr. J. Keepers of the biophysical chemistry group at Rutgers; the photographs were obtained by Professor P. Orenstein of the Department of Visual Arts at Rutgers. We acknowledge their kind assistance.

**Chemical Principles in the Laboratory, Third Edition**
Copyright © 1985 by Harper & Row, Publishers, Inc.

All rights reserved. Printed in the United States of America. No part of this book may be used or reproduced in any manner whatsoever without written permission, except in the case of brief quotations embodied in critical articles and reviews. For information address Harper & Row, Publishers, Inc., 10 East 53d Street, New York, NY 10022.

**Library of Congress Cataloging in Publication Data**

Bryan, Robert F.
    Chemical principles in the laboratory.

    Rev. ed. of: Chemical principles in the laboratory/
Lester R. Morss.
    Includes bibliographical references.
    1. Chemistry—Laboratory manuals.   I. Boikess,
Robert S.  II. Morss, Lester R.   Chemical principles
in the laboratory.  III. Title.
QD45.B889 1985       542        84-27921
ISBN 0-06-040814-6

# Contents

*Introduction*    1
*Check in and Preliminary Preparation*    5

**Experiment**

1. Separation, Estimation, and Identification of the Components of a Ternary Mixture    29
2. Chromatographic Separation and Identification of Metal Ions in Aqueous Solution    63
3. Determination of the Empirical Formula of a Metal Sulfide    77
4. Synthesis and Purification of Potassium Bromide and Bromate    93
5. Determination of the Standard Molar Volume and the Gas Constant, $R$    113
6. Determination of Molar Mass for a Volatile Liquid by Dumas's Method    131
7. Determination of the Heat of Formation of Magnesium Oxide    149

8  Evaluation of Avogadro's Number from the Measured Density and X-Ray Diffraction Data for a Crystalline Organic Compound   171

9  Water Analysis   189

10  Ionic Interactions and Qualitative Analysis   209

11  Molar Mass Determination by Freezing Point Depression   227

12  The Solubility Product of Calcium Iodate   243

13  Determination of an Equilibrium Constant by Spectrophotometry   259

14  Acid-Base Equilibria
    Part I   The Use of Indicators to Determine pH   277
    Part II  Buffer Solutions and the pH Meter   301

15  Introduction to Volumetric Analysis: Acid-Base Titrations in Aqueous Solution   323

16  Determination of Antacid Action   341

17  Oxidation-Reduction Reactions of Vanadium in Acid Solution   363

18  Synthesis of a Potassium Oxalatoferrate Complex   377

19  Standardization of a Potassium Permanganate Solution   387

20  Determination of the Empirical Formula of an Oxalatoferrate Complex by Permanganate Titration   401

21  Kinetics of the Oxidation of Iodide Ion in Acidic Solution by Hydrogen Peroxide   417

22  Synthesis of Some Polymers   435

**Appendix**

Design, Use, and Care of Volumetric Glassware   455

# Chemical Principles
in the Laboratory

# Introduction

## WHAT LABORATORY WORK IS ABOUT

The third edition of Boikess and Edelson's *Chemical Principles* opens with the statement:

> Science is an organized body of knowledge based on a unique method of looking at the world. Any explanation of what is seen is based on the results of experiments and observations that must be verifiable by anyone who has the time and means to repeat them.

*Chemical Principles in the Laboratory* is your guide to testing this statement. It is a collection of 22 experiments in general chemistry designed to illustrate and amplify the theoretical material in the main text. The experiments do not break any new ground in chemistry—you cannot hope to do that kind of work until your training in chemistry is complete. But their value is threefold for they will

1. Show you the kinds of things chemists do to test a theory.
2. Introduce you to the idea of making accurate measurements and to recognizing their power and limitations.
3. Help you learn and perfect simple laboratory techniques.

Let's say a word about each of these topics. First, certain areas of chemistry are more amenable to simple experiment than others. For example, masses and volumes can

## 2  INTRODUCTION

be measured with fair accuracy using simple apparatus. Carrying out subtle experiments in atomic and molecular structure, however, involves complex and expensive equipment not often available to the first-year student. That fact is reflected in the choice of experiments in this manual and others like it. Some chapters of the main text have more experiments to illustrate their content than others, but the kind of relationship between theory and experiment that is emphasized is quite general.

You will benefit most from your course in chemistry if you adopt the simple plan of reading this manual and the main text together. To make that easier for you, each experiment is accompanied by a special box headed Associated Readings. In the box are references to the sections of *Chemical Principles* that the experiment is designed to illustrate. You should read those sections of the main text before coming to the laboratory. That will help you understand the experiment better. You should then reread them after the laboratory work. That will help you understand the theory better.

Now for the second point. In all the experiments you are expected to get results as accurate as techniques allow. In most experiments your answers can be matched against expected values so you can see how well you have carried out the work. Poor agreement between theory and experiment is not going to overthrow any chemical theories—not at this stage in your chemistry career—but always remember when you have become a competent experimentalist that just such poor agreement may set you on the road to discovery.

That brings us to the third point. To get good answers you need to be a good experimentalist. If you have never worked in a chemistry laboratory before, possibly your first efforts will be clumsy. We are all of us awkward when we begin to learn a new skill. How long did it take you to become a decent tennis player or swimmer? The key to success in experimental work is the same as that in any other task—practice. The first time you step into the laboratory you will be confronted by new techniques and equipment, but you will be given the chance to practice your basic skills before they are put to the test for graded work. As you use those techniques over and over again, you will gradually acquire confidence in your abilities, perfect your skills, and experience the deep satisfaction that comes from doing something really well.

In the early experiments you will find that laboratory procedures are spelled out in fair detail. As you progress, you will notice you are left a bit more on your own. We think that's the way it should be and hope you will, too. Remember that if you run into difficulty, your instructor is always there to answer questions.

There is one more thing to be said about accurate results. No measurement is ever perfect. We introduce you to the idea of recognizing the uncertainty inherent in the measurements you make and of calculating the effect of that uncertainty on the accuracy of your final answers. We do not give a formal treatment of such things as accuracy versus precision. Rather, we invite you to estimate the uncertainty in your measurements and ask that you assess the effects in each case. We believe that making you do a little thinking for yourself is a good thing. Once again, in case of difficulty, don't hesitate to ask your instructor for help.

## HOW TO USE THIS MANUAL

The manual contains all the printed material you will need for the laboratory part of your course. Besides the manual you should bring the following items to the laboratory with you on opening day:

1. A ballpoint pen for recording data. Choose a pen that does not smear.
2. A sheet of carbon paper for making duplicate entries of your data.
3. A No. 3H or harder pencil for plotting entries on graphs.

# INTRODUCTION

4. A laboratory coat or apron and proper clothing. We discuss this in more detail in the section called Check-in and Preliminary Procedures.
5. A calculator. Strictly speaking, all you need is one that can add, subtract, multiply and divide. However, a model with a log function will be useful for pH calculations and a programmable model will be useful in Experiment 14.

Each experiment in the manual is presented in six sections. These are discussed below.

## Description of the Experiment to be Performed

This section contains a statement of the aim of the experiment, a list of equipment, reagents and other materials required, and sufficient background information to illuminate both the theory involved and the practical procedures to be followed. Be sure to use the Associated Readings to help you focus your reading of the main text. Of particular importance are the Safety Notes, which highlight potential hazards and how to avoid or treat them.

## Prelaboratory Preparation Sheet

An unprepared student is a danger in the laboratory. Before coming to the laboratory, you are expected to complete a prelab sheet of questions based on the theory and procedures associated with the experiment. You will not be allowed to participate in the laboratory work unless you present a conscientiously prepared prelab sheet.

## Data Sheets

Data sheets are provided in duplicate for each experiment. Always enter your observations and measurements directly onto the sheets, using a pen. Use carbon paper to make duplicate entries. Do not make separate entries on the two sheets. Always provide the units associated with each quantity; e.g., mL, g, s, mmHg, K, etc. If you make an error in a data entry, strike out the incorrect entry with a single line but leave the original entry legible. Make the new entry directly above the old.

When two of you are collaborating on an experiment, one of you should do the data entries for both, and do them at the same time, but each of you should check the correctness of the entries made.

Never use pieces of scratch paper to record weighings and other measurements with a view to entering them later on your data sheet. This is wasteful of time and is bad technique. Such scraps tend to get lost, and you may forget what the numbers mean and enter them in the wrong place on the data sheet. Cultivate the habit of neatness in data entry; your grades will reflect it.

Give the top copy of your data sheet to your instructor at the end of the laboratory period. The instructor will initial the bottom copy and you keep that copy as proof of your attendance. Never alter a data sheet after leaving the laboratory without at once explaining the circumstances to your instructor. The instructor's copy is the copy of record.

## Report Sheet

Each report sheet contains questions or entries designed to interpret and extend the observations made during the experiment. In some cases the data and report sheets are combined. As with the other sheets, use a pen and be neat. Reports are normally due at the start of the next laboratory period, but some are to be completed at the end of the experiment itself. Check with your instructor if you are in doubt when a report is due.

## Worksheets

Where a report requires extensive numerical calculations, do them on the worksheet. Many of the reports contain spaces for arithmetic work and you may set out your work there. However, you should never present a result on your report sheet without evidence of how it was obtained numerically. Your arithmetic work must appear on either the report sheet or the worksheet. Where you use the worksheet, staple it firmly to your report and submit both together.

## Stockroom Slips

For most of the experiments, you will need specialized equipment over and above the contents of your locker. When that equipment is to be checked out of the stockroom, you will need a stockroom slip. These are found at the back of the manual. Tear out the perforated slip for the experiment in question, write your name on it legibly, and present it to the stockroom clerk to obtain the equipment needed. All equipment checked out of the stockroom must be returned by the end of the laboratory period. Do not wait until the very end of the period, but clean and return stockroom items as soon as you have finished with them. Be sure to collect your signed slip from the stockroom clerk. You will be charged for all equipment listed on signed slips held by the stockroom at the end of the semester.

# Check-in and Preliminary Preparation

### AIMS

1. To check the contents of your locker
2. To familiarize yourself with laboratory safety equipment and procedures
3. To familiarize yourself with the location and use of communal equipment
4. To clean all glassware in preparation for future work
5. To learn the use of the Bunsen burner
6. To learn techniques of handling glass tubing and to prepare an aspirator trap bottle
7. To learn the use of the chemical balance and to weigh out the sample to be analyzed in the first experiment
8. To learn the use of the Buchner filter funnel and flask

### EQUIPMENT

Locker key and towel
All apparatus on your list of issued equipment
Other communal laboratory apparatus and equipment
Glass tubing and files
Aluminum foil and rubber bands

## REAGENTS AND OTHER MATERIALS

Cleaning materials (detergent, soap powder, or solution)
Various unknown mixtures for later analysis
Glycerin lubricant
Sand and water

## BACKGROUND TO THE FIRST DAY

The opening day's operations introduce you to the chemical laboratory, to the equipment you will use, and to certain basic operations and procedures you will carry out in later experiments. You will also meet the instructor in charge of the section to which you have been assigned. You are expected to follow the instructor's directions promptly on all matters relating to laboratory procedures and safety. In turn, your instructor will answer questions you may have and advise you how to carry out new and unfamiliar tasks when required.

You should now read through the material in this section and then complete the prelaboratory preparation sheet on page 17. That sheet should be presented to your instructor at check-in time.

## PROCEDURE

### ABOUT SAFETY NOTES

Each procedure section for the various experiments will begin with a set of safety notes. Always read them carefully and follow the directions they contain. Remember that in the laboratory you are in unfamiliar surroundings and are asked to perform unfamiliar tasks. Both these conditions make the chance of an accident happening greater than normal. Be alert for potential hazards, follow prescribed directions carefully, and, above all, use your common sense. Think ahead. If some action doesn't seem particularly safe, it usually isn't. So don't do it! If you are unsure about the proper way to carry out some operation, ask your instructor for advice.

Today you will be introduced to the location and use of laboratory safety equipment. You will also be instructed in personal safety in the laboratory, including the importance of suitable clothing and eye protection. Begin today to *acquire safety consciousness.* Remember that you must be concerned not only with what you are doing but with what others may inadvertently do to you.

Report to your assigned laboratory room at the scheduled time. Your instructor will meet you and provide you with directions on how to obtain a key or combination for your locker and a towel for drying your glassware. Guard the key or combination carefully, bringing it to laboratory with you for each session.

On this opening day you will carry out the operations described below, though not necessarily in the sequence indicated.

### Checking the Contents of Your Locker

Your locker contains most of the equipment you will use on a regular basis in the course of the semester. Your instructor will provide you with a list of the locker contents. Open your locker, find your safety goggles, and put them on. This should become your regular practice as soon as you enter the laboratory, and you should continue to *wear your goggles*

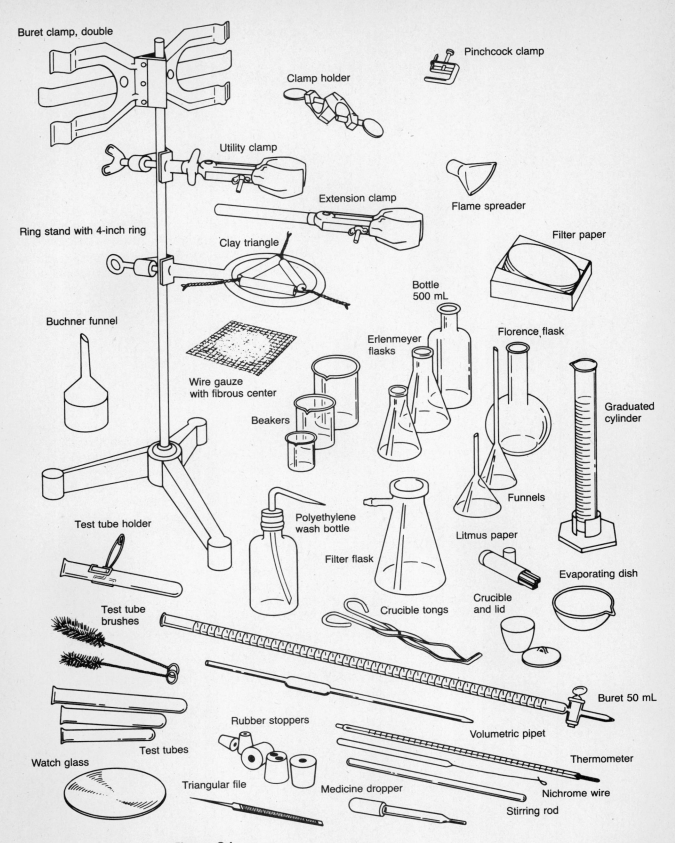

**Figure C.1**
Some typical pieces of chemical equipment.

*at all times when in the laboratory.* Your instructor will advise you of any special requirements concerning the wearing of contact lenses, etc., in the laboratory.

Check the equipment in your locker against the list provided. Have any missing or damaged items replaced as your instructor directs. Be sure that all glassware is free from chips and cracks. Typical pieces of apparatus are illustrated in Figure C.1.

When you are satisfied that all of the pieces of equipment are in good condition, sign the locker receipt and return it to your instructor. From now on you are responsible for any breakages or loss or damage of equipment. Your instructor will explain the procedure for replacing lost items.

## Laboratory Safety Equipment and Procedures

Your instructor will issue you a sheet outlining safety rules that apply to your laboratory, and will explain the importance and significance of them. Please read the sheet carefully and keep it with you for future reference. You may be asked to sign the statement at the foot of the sheet indicating that you have done so, and your instructor may hold a safety quiz.

Your instructor will point out to you the locations of various pieces of safety equipment in the laboratory and explain their use. Among these items should be some or all of the following:

Safety showers
Eye-wash fountains or bottles
First-aid kits
Fire extinguishers
Fire blankets
Fire alarm
Emergency telephone

Note the locations of these items on your data sheet.

Your instructor will also point out to you and explain the posted procedures to be followed in case of an emergency involving personal injury. Study these carefully at this time so that you will know at once what needs to be done should the need arise.

Your instructor will also explain the procedures to be followed in case of a fire at your bench, in your laboratory, or elsewhere in the building. Be sure that you fully understand what is to be done in each case.

## Location and Use of Communal Equipment

Certain pieces of equipment are kept in the laboratory for communal use. Typically these include:

Burners and strikers
Rubber tubing
Buret stands
Ring and tripod stands
Rings and clamps
Test tube stands
Test tube brushes
Cleaning materials

Your instructor will tell you where these items are and will demonstrate their use. Note the proper locations for each item on your data sheet for future reference. Treat all communal equipment carefully, always cleaning it and returning it to its proper place after each use.

## CLEANING GLASSWARE

To obtain good experimental results all glassware used in experiments must be scrupulously clean. You should therefore begin the laboratory course with your glassware clean and should maintain it in that condition. Never leave the laboratory at the end of the day without having cleaned your equipment. Many materials which are readily cleaned up in the course of the day set fast if neglected and become very difficult to remove if left unattended even for a few hours.

Partially fill a sink with hot water and add a small amount of the detergent or other cleaning material provided. Two students may comfortably share a sink. Use a sponge to wash your beakers and other flatware and use a test tube brush to clean flasks and test tubes. Rinse the glassware with hot water and use the towel to dry it. Test tubes should be inverted in a test tube stand and allowed to drain dry. When dry, replace the glassware in your locker. If it is communal property, return the test tube stand to its proper place.

Use a wet sponge to rinse off the top of your laboratory bench. Squeeze the sponge dry over the sink and use it to dry off the bench top. Go through this bench-cleaning process at the end of each laboratory period, more frequently during experiments involving a lot of bench work. Always keep your bench top clean and tidy.

## USE OF THE BUNSEN BURNER

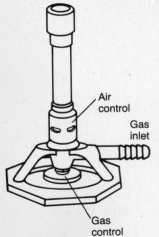

**Figure C.2**
A typical design of bunsen burner.

The Bunsen burner is used for most heating jobs in the general chemistry laboratory, but it is never used to heat volatile or flammable liquids. It burns a mixture of gas and air and when properly regulated can produce a wide range of flame sizes and heat.

The burner has two controls. The knob at the base controls a needle valve that regulates the supply of gas to the burner. The perforated sleeve on the stem of the burner regulates the amount of air in the mixture. A typical design is shown in Figure C.2. For proper use, the amounts of both air and gas must be correctly regulated.

To set up the burner, attach one end of a length of rubber tubing to the gas inlet on the burner and the other to the gas tap on the bench. Completely close both the air and gas control on the burner and fully open the bench gas tap. Open the gas control on the burner about half a turn and ignite the gas at the top of the stem with a striker. The flame will be yellow and if used in that condition will deposit soot on any cold surface with which it comes in contact. Adjust the gas control on the burner until a flame of the desired height is obtained and then open the air control sleeve until a clear blue flame with a well-defined blue-green central cone is obtained. If too much air is admitted the flame will be unstable and noisy; always use a quiet clear flame.

The central blue cone consists of unburned gas and is quite cool. The hottest part of the flame is about 1 cm above the tip of the central cone. The Bunsen burner has a wide range of flame sizes, depending on how much gas is admitted. Always choose the smallest flame size needed for the job at hand. Don't use a vigorous 10-cm flame to heat 5 mL of water in a test tube, for example. Such a task needs only a gentle 3-cm flame.

Test tubes, never more than about a quarter full of liquid, may be heated directly in a gentle burner flame by moving them into and out of the flame, playing the flame over all parts of the glass in contact with liquid so as to avoid local overheating.

The burner should not be applied directly to any other piece of glassware containing solid or liquid. Beakers and flasks containing liquid should always be supported on a tripod or ring stand and always with a piece of wire gauze between the vessel and the flame. The gauze distributes the heat from the burner evenly over the base of the reaction vessel and averts the risk of the glass cracking or shattering through local overheating. Figure C.4 shows typical arrangements for heating liquids. Choose carefully the size of the beaker or flask to be used. No vessel should normally be more than half full and never more than three-quarters full.

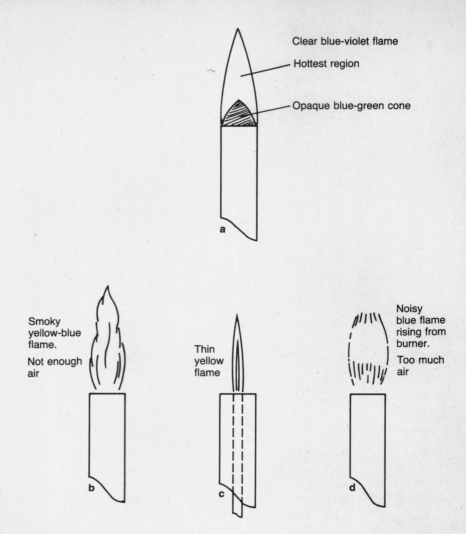

**Figure C.3**
Burner flames
a Properly adjusted
b Insufficient air
c Back-burning. Extinguish and relight using less air
d Too much air

Remove small beakers or flasks containing hot liquids from the stands by allowing them to cool for a minute or two and then placing your fingers around the topmost outer surface of the vessel and lifting it clear of the stand. If the directions given in the last paragraph concerning the volume of liquid in relation to the volume of the container are followed, then the rim of the vessel will be cool enough to handle within a few seconds of removing the burner flame. Never use crucible tongs to handle beakers or flasks. They are not designed for this purpose and a spill is inevitable if you do.

### Handling Glass Tubing

Your instructor will demonstrate the proper techniques for cutting, heating, and bending glass tubing. The following notes may be useful in supplementing the demonstration. Always be sure that your bench top is clean and dry before working with glass tubing.

### Cutting Tubing

Place the length of tubing to be cut on the clean, dry bench top. Grip a triangular glass file in your hand so that a flat side is uppermost. Fully extend your forefinger along the flat surface. Steady the tubing with your other hand and place the near edge of the file on the glass tube at the point of the desired cut. Pressing down firmly with your forefinger, draw the edge of the file sharply across the tubing so that the whole length of the file passes across the glass. This process should have produced a sufficiently deep scratch on the glass. Avoid ''sawing'' the glass with the file.

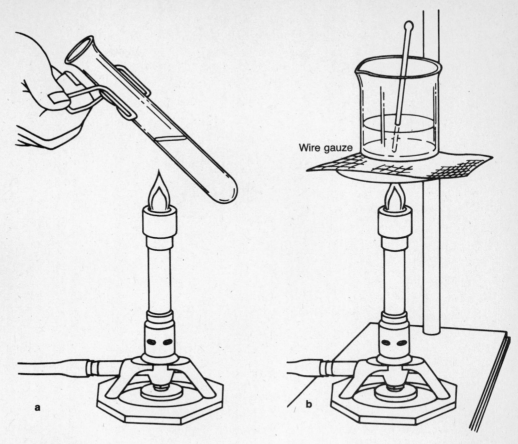

**Figure C.4**
Using the burner to heat liquids or solutions.
**a** Heating liquid in a test tube. Don't overload the tube and aim the burner immediately below the top of the liquid rather than at the base of the tube. Use a small clear flame and avoid pointing the mouth of the tube at your neighbor in case of spurting.
**b** Heating liquid in a beaker. Always use a wire gauze between the base of the beaker and the burner flame so as to distribute the heat evenly. Match the beaker size and contents so as to avoid over or under-filling. A stirring rod or boiling chips may be used to prevent bumping arising from superheating of the liquid.

Grip the glass tubing in both hands, as shown in Figure C.5, so that the scratch is pointing away from you and your thumbs are positioned on the other side of the tubing just opposite the scratch mark. Gently pull on the tubing while pushing toward the cut with your thumbs. A quite gentle pressure of this sort should cause the glass tube to break cleanly at the point of the scratch mark. Don't exert a lot of pressure if the glass does not break readily, but repeat the filing step at the same place on the tube. The tubing should break cleanly without any jagged edges. If the ends are not clean it is best to try again with a fresh piece of tubing. To reduce the risk of injury, never work with tubing having jagged ends.

### Bending the Tubing
Fit a fishtail adapter to the top of the Bunsen burner so as to spread out the flame. Light the burner in the way described in the previous Section to obtain a clear blue flame about 3 cm high. Grip one end of the tubing to be bent in each hand and place its center in the flame. Twirl the tube around its long axis so as to heat it evenly. Continue heating until the glass begins to soften, remove it from the flame, and gently bring each end together so as to produce a bend. A good bend will show little or no constriction of the bore at the bend. Set the tubing aside to cool.

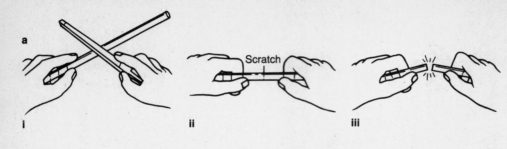
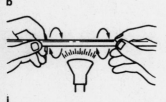 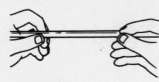  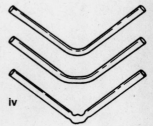

**Figure C.5**
**a** Steps in cutting glass tubing.
  i  make a single scratch with a triangular file
  ii place the thumbs together opposite the scratch
  iii pull and bend quickly
**b** Steps in making a bend in glass tubing
  i  Rotate the tubing in the flame until it begins to soften
  ii Remove tubing from flame and wait two seconds
  iii Firmly push ends together to make desired bend
  iv Top: a good bend free from constrictions
     Middle: a poor bend, not enough heat
     Bottom: a poor bend, too much heat causing collapse

### Fire-polishing
The tube ends produced by cutting should be clean, and they will be sharp. Before incorporating the tubing into any piece of apparatus, you must fire-polish the sharp edges to round them off so as to reduce the risk of cuts and to make it easier to fit the tubing into stoppers, etc. This is done by simply placing the end of the tube in the burner flame and rotating the tubing until the softening of the glass removes the sharp edge. Do not overheat the ends, as they may collapse and cause constrictions.

### Preparing an Aspirator Trap Bottle
The sinks in the laboratory are fitted with aspirator taps to allow suction filtration with the Buchner funnel and flask. To avoid water from the aspirator backing up into the filter flask, a trap bottle is inserted in the line between the tap and the filter flask. A typical trap bottle is shown in Figure C.6 and may be constructed using an Erlenmeyer flask.

Cut two 25-cm lengths of glass tubing and make a right angle bend in the center of each. Allow the tubes to cool and cut the ends to size so as to make an entry and exit tube. Remember that glass is mechanically quite fragile, so that the tubes should be no longer than is required. They should not project more than 1 cm above the top of the apparatus and the horizontal arms should be no more than 4 cm long. Use the figure as a guide in deciding where to cut the tubing.

Fire-polish the ends of the tubes and allow them to cool. Select a suitable flask and two-holed stopper. Lubricate the stopper holes with water or a drop of glycerin and insert the two tubes into the stopper. Be careful not to use undue force in inserting the tubing into the stopper and always grip the tubing as close to the stopper as possible.

Secure the trap bottle on a ring stand with a clamp. Take a length of aspirator tubing and wet each end with tap water. Attach one end firmly to the aspirator intake and the other to the exit tube of the trap bottle. Place your largest beaker under the water

**13** Check in and Preliminary Preparation

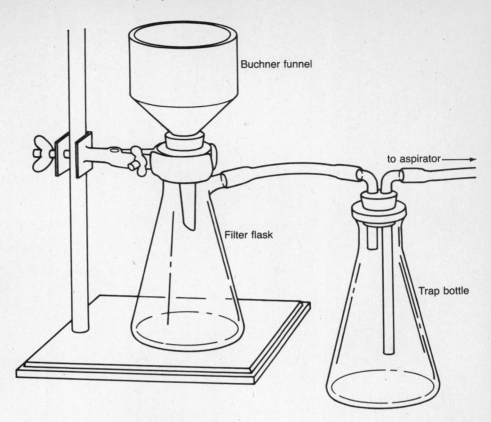

**Figure C.6**
Buchner filter funnel and flask with trap bottle. The filter flask should be secured in a stand and clamp.

outlet of the aspirator, so as to minimize splashing, and open the aspirator tap fully. Test the suction developed at the entry tube of the trap bottle by placing your finger over the end of the tubing. If the suction is weak or absent, check that the stopper fits well and is properly seated in the neck of the flask. If this does not help, call your instructor to check the aspirator itself. If a good suction is obtained, turn off the aspirator tap and disconnect the trap bottle, storing it in your locker until you carry out the practice filtration. Return the aspirator tubing to its proper location.

## Use of the Balances

Many varieties of chemical balance are in use in different laboratories and each involves different operating directions. Your instructor will provide you with detailed operating directions for the balances in your laboratory, and will demonstrate their use. Study these directions carefully and watch a demonstration before trying to use a balance on your own.

Fine balances, designed to weigh to the nearest milligram or better, are very sensitive instruments, costing several thousand dollars each. Treat the fine balance with the greatest respect, carefully following the operating directions and never forcing any of the controls. Never add chemicals or reagents to any container at the fine balance. Spills of such materials could find their way into the inner workings of the balance and lead to corrosion or mechanical damage.

Depending on their design, rough balances are capable of weighing objects to the nearest 10 or 50 mg. These balances should be used when weighing out chemicals. If only an approximate mass is required, then these balances will suffice. If a mass to the nearest milligram is required, weigh out the approximate amount of reagent into a suitable container on the rough balance and then determine the exact mass on the fine balance. In the experiments described in this manual, all exact masses of chemicals are found by determining the difference between the mass of the container + reagent and that of the empty container *after* the reagent has been transferred. There is therefore no need to add chemicals directly at the fine balance.

When you have satisfied your instructor that you have mastered the use of the balance, he will assign you an unknown mixture, which you will analyze in the first experiment. Record the sample number of the mixture on your data sheet.

### Weighing Out the Sample

Take a clean, dry 100-mL beaker and establish its mass with the rough balance. Gently tip about 6–7 g of the assigned mixture into the beaker directly from the sample bottle.

Replace the lid of the bottle and on your data sheet record the combined mass of beaker and reagent to the best accuracy allowed by the balance. Replace the sample bottle on the reagent shelf.

The techniques used in the separation experiment do not justify the use of the fine balance—and a measurement to the nearest 50 mg will suffice—but with the permission of your instructor you may use the weighed sample in its container to practice the use of the fine balance.

Cut an 8-cm square of aluminum foil and with it cover the top of the beaker. Secure the foil to the sides of the beaker with a rubber band so as to make a tight seal and store the beaker and its contents in a clean safe place in your locker until the next laboratory session.

### Using the Buchner Funnel and Flask

Your instructor will demonstrate the use of the Buchner filter funnel and flask. The apparatus is illustrated on page 13 in Figure C.6. It is a device used to filter liquids from solid-liquid mixtures and does so by creating a partial vacuum in the filter flask so

**Figure C.7**
Dispensing a solid reagent from its container.

that the pressure of the air drives the liquid through a filter paper into the flask, leaving the solid behind in the filter bed. The liquid in the flask is called the **filtrate** and the material remaining behind in the filter is called the **residue**.

The apparatus should always be secured by a stand and clamp as shown in Figure C.6. For successful filtration a number of points must be observed:

1. The aspirator must be turned on and a good suction must be present before filtration begins.
2. The proper filter paper must be used. It must lie flat on the perforated base of the funnel, be moistened with the solvent in use in the mixture, and not be allowed to dry out during filtration; otherwise solid will pass under the paper and into the filtrate.
3. Liquids being decanted into the funnel should always be run down a glass stirring rod onto the center of the filter paper, and the rod should finally be touched against the inside surface of the funnel to ensure that the last drop of liquid is transferred.
4. If the filtrate is to be retrieved, a trap bottle should be inserted in the line to prevent backflow from the aspirator into the filtrate.

Practice filtration techniques, working in pairs. Check out one filter funnel and flask between you and set it up, connecting it to the aspirator. Take a 20-cm test tube and add to it about 3 cm of sand. Half fill the tube with tap water, stopper it, and shake well to mix the contents. Separate the two components of this simple mixture by filtration, observing the precautions given above. Have your instructor check your performance and then return the filter apparatus to the stockroom.

## Final Tasks

When you have finished the day's work, return any equipment you may have checked out to the stockroom. Clean and return all communal equipment you have been using to its proper location. Clean and dry any glassware used, and return your equipment to the locker. Equipment left on the benches or other parts of the laboratory will be placed in a common pool.

Rinse off and dry your bench top. Complete any missing items on your data sheet and take it to your instructor for his inspection. When the instructor is satisfied with your performance you will be given permission to leave the laboratory. Return to your locker, remove and store your safety goggles, and depart. This sequence of operations should be followed at the end of each laboratory session. It is particularly important that you continue to wear your safety goggles up to the moment of leaving the laboratory.

Name _____  Instructor _____

# PRELABORATORY PREPARATION SHEET

1. Before checking into the laboratory, you should have been assigned a locker or station in one of the laboratory rooms used in the course. Enter that information here.

    Laboratory room number _____ Locker number _____

2. Certain items of protective gear and clothing are required for laboratory work. List them below.

   _____
   _____
   _____

3. Why do you suppose you are required to wear safety goggles at all times while you are in the laboratory?

   _____
   _____

4. Why is it important to know in advance what to do in case of an emergency involving personal injury or fire?

   _____
   _____
   _____

5. Why must the glassware used in chemical experiments be clean?

   _____
   _____

6. How is the size of the flame on a Bunsen burner controlled?

   _____
   _____

7. Why are chemicals never weighed out directly from the reagent bottle at the fine balance?

   _____
   _____

8. Why should the ends of glass tubing used in experiments be fire-polished?

   _____
   _____

9. Why should you keep your bench top clean and tidy at all times?

   _____
   _____

**10.** Why should you take special care to keep communal equipment in good order?

_____

_____

Name _____  Instructor _____

## DATA SHEET

1. Check off the following items as you complete the work associated with each. The list will serve as a guide to the tasks you should perform.

   Equipment list collected from assistant _____

   Locker key and towel obtained _____

   Locker located and safety goggles found _____

   Safety goggles donned _____

   Locker contents checked _____

   Missing or defective equipment replaced _____

   Equipment list signed and returned _____

   Glassware cleaned and dried _____

2. Enter Location of Safety Equipment:

   Eye-wash fountain or bottle _____

   Safety shower _____

   Fire extinguishers _____

   _____

   First-aid kit _____

   Fire alarm _____

   Emergency procedures _____

   Emergency telephone _____

   Other (specify) _____

   _____

3. Enter Location of Communal Equipment:

   Burners and strikers _____

   Ring stands _____

   Rings and clamps _____

   Tripod stands _____

   Buret stands _____

Name _____    Instructor _____

## DATA SHEET

1. Check off the following items as you complete the work associated with each. The list will serve as a guide to the tasks you should perform.

   Equipment list collected from assistant _____

   Locker key and towel obtained _____

   Locker located and safety goggles found _____

   Safety goggles donned _____

   Locker contents checked _____

   Missing or defective equipment replaced _____

   Equipment list signed and returned _____

   Glassware cleaned and dried _____

2. Enter Location of Safety Equipment:

   Eye-wash fountain or bottle _____

   Safety shower _____

   Fire extinguishers _____
   _____

   First-aid kit _____

   Fire alarm _____

   Emergency procedures _____

   Emergency telephone _____

   Other (specify) _____
   _____

3. Enter Location of Communal Equipment:

   Burners and strikers _____

   Ring stands _____

   Rings and clamps _____

   Tripod stands _____

   Buret stands _____

Test tube stands _____

Test tube brushes _____

Rubber tubing _____

Cleaning materials _____

Other (specify) _____

_____

4. Demonstrations:

Weighing on fine balance _____

Weighing on rough balance _____

Burner handling _____

Glass handling _____

Buchner filtration _____

5. Activities:

Trap bottle prepared _____

Filter apparatus obtained _____

Filtration done _____

Filter apparatus returned _____

Sample obtained and weighed _____

End of lab clean up _____

Sample number _____

Mass of sample + beaker _____

Test tube stands _____

Test tube brushes _____

Rubber tubing _____

Cleaning materials _____

Other (specify) _____

_____

4. Demonstrations:

Weighing on fine balance _____

Weighing on rough balance _____

Burner handling _____

Glass handling _____

Buchner filtration _____

5. Activities:

Trap bottle prepared _____

Filter apparatus obtained _____

Filtration done _____

Filter apparatus returned _____

Sample obtained and weighed _____

End of lab clean up _____

Sample number _____

Mass of sample + beaker _____

# EXPERIMENT 1

# Separation, Estimation, and Identification of the Components of a Ternary Mixture

### AIM

To separate the three components of a ternary mixture on the basis of differential solubility, to identify each component, and to establish the percentage composition of the mixture by mass.

### EQUIPMENT

| | |
|---|---|
| From stockroom: | Buchner funnel and flask |
| From locker: | Two 250 mL, two 100 mL, two 50 mL beakers |
| | Test tubes |
| | Watchglass |
| | Trap bottle |
| | Thermometer |
| | Stirring rods |
| | Rubber policeman |
| Laboratory: | Burner |
| | Stand and clamps |
| | Wire gauze |

Nichrome wire
Melting point tubes
Aspirator tubing
Filter paper
Cobalt glass
No. 0 and No. 5 rubber stoppers
Centrifuge and centrifuge tubes
Rubber bands

## REAGENTS

Solvents for extraction: Dichloromethane and distilled water

Reference reagents for ion analysis: $0.1M$ solutions of $Co(NO_3)_2$, $CuSO_4$, $Fe(NO_3)_3$, $NiCl_2$, $NH_4Cl$, $NaCl$, $KBr$, $KI$, $Na_2CO_3$, $KNO_3$, $Na_2SO_4$, $AgNO_3$, $BaCl_2$; solid $NaCl$, $KCl$, $CaCO_3$

Supplementary reagents: $1M$ solutions of $KSCN$ and $NH_4SCN$, solid $NaF$, 1% dimethylglyoxime in alcohol, diphenylamine-$H_2SO_4$ solution, lime water, acetone for drying

Acids and bases: Concentrated $HNO_3$ and $H_2SO_4$, $3M$ $HCl$ and $HNO_3$, $4M$ $NaOH$ and $NH_4OH$ (aqueous ammonia)

## BACKGROUND

### Associated Readings

To help broaden your understanding of Experiment 1, we suggest that you read the following sections of the third edition of Boikess and Edelson's *Chemical Principles*, before coming to the laboratory. These sections provide additional information on the topics treated in the background material.

You may also find it helpful to reread the text sections after you have completed the experiment, as it is an appreciation of the interplay between theory and experiment that leads to the deepest understanding of chemistry.

**Chapter 1** The Science of Chemistry
       **1.1** The Nature of Science: The Scientific Method
       **1.2** Chemistry as a Physical Science
**Chapter 2** The Language of Chemistry
       **2.2** Chemical Formulas
       **2.3** Chemical Names and Formulas
       **2.4** Balanced Chemical Equations
       **2.5** Important Types of Chemical Reactions
       **2.6** States of Substances

A mixture is an aggregate of two or more substances each of which retains its chemical identity and does not react chemically with any of the others. In chemical syntheses or in commercial extraction processes such as mining, few substances are obtained in their pure state; most more frequently occur as mixtures. Separation of the components of a mixture is thus one of the commonest tasks performed by chemists. Many different techniques have been developed for separation based on the various and differing physical properties of the components of the mixture. In this experiment we will make use of the observation that different substances do not dissolve in a particular solvent to the same extent. This technique, known as differential solubility, is quite widely used as, for example, in the extraction of caffeine from coffee beans with liquid carbon dioxide. The caffeine is removed in the $CO_2$, but the remaining alkaloids and flavor molecules do not go into solution.

The mixture to be analyzed in this experiment contains one organic and two inorganic components. The organic component is soluble in a variety of organic solvents—we will use dichloromethane—but is effectively insoluble in water. One of the two inorganic components is a metal carbonate insoluble in water, the other is a water-soluble salt. Neither inorganic component is soluble in organic solvents to any measurable extent.

We will proceed by first extracting the organic component with dichloromethane and separating the resultant solution from the two inorganic solids by suction filtration. The water-soluble inorganic component will then be separated in the same way from its insoluble partner, using water as a solvent.

To estimate the quantity of the organic component we will recover it from solution by evaporation of the solvent and find its mass by weighing. The mass of the insoluble inorganic component will be found by weighing after drying, and the mass of the water-soluble material will be estimated by difference.

The organic component will be identified by its melting point, a characteristic physical property. The identification will be confirmed by carrying out a mixed melting point determination. When two separate pure samples of the same material are mixed together, the mixture has the same melting point as either pure sample. By contrast, when samples of two different substances having the same melting point are mixed the melting point of the mixture is not the same as that of the individual components but is lowered.

The ions present in the aqueous solution of the water-soluble salt will be identified by applying some simple qualitative analytical tests. You will carry out a complete set of standard tests on reference samples, noting your results, and repeat the tests on your sample of unknown until the constituent ions have been identified. Note that some salts, e.g., ferrous ammonium sulfate, $Fe(NH_4)_2(SO_4)_2 \cdot 6H_2O$, contain more than one cation. Salts containing two different anions are much less common. You will be told if you should test for the presence of more than one cation or anion.

A small amount of the insoluble metal carbonate will be brought into solution by reaction with $3M$ nitric acid, $HNO_3$, and you will repeat the same tests to establish the identity of the single cation present. It is worth bearing in mind that ammonium carbonate and the carbonates of the alkali metals are soluble in water but that all other metal carbonates are insoluble.

The experiment will occupy two laboratory periods. In the first you should separate the three components and isolate the organic component, and you should carry out as many of the qualitative analytical tests on your water-soluble unknown as time permits. In the second period you will weigh the organic and water-insoluble fractions, determine the identity of the organic component, and complete the analytical tests on each inorganic component. Fill out the data sheets as you go along, recording your observations as they are made. Show your data sheet to your instructor at the end of the first period but do not detach it at that time. Your instructor will initial the sheet to verify your attendance. Hand in the completed data sheets and report sheet at the end of the second period.

## PROCEDURE

### SAFETY NOTES

For many of you this will be your first serious encounter with a variety of chemical reagents. Some of these substances are caustic or corrosive and must be handled with great care. Others give off noxious vapors; still others are poisonous if ingested. However, all may be safely handled and used in the experiments you are called on to perform if you follow the instructions provided. Please obey the general rules outlined below as they apply to this and other experiments.

## Strong Acids and Alkalis

Reagent acids and alkalis in the laboratory are of varying strengths and concentrations. The very highest concentrations of acid or base are referred to as **concentrated.** Bottles or other containers holding these reagents should be kept in plastic trays in a designated fume hood and are not to be removed from the hood. When you require some of these reagents, take the container in which they are to be placed to the hood and make the transfer at the hood. Pour from the bottle over the plastic tray (or use the dispensing pipets provided) so that any spills land in the tray. Always clean up any spilled reagent. Carbonate or bicarbonate will be found next to the trays and should be used to neutralize any sizable acid spill. Use a thoroughly wet sponge to wash the neutralized spill down the drain at the rear of the hood and flush the drain with cold water. Because the vapors are toxic or corrosive, and because of the risk of spillage, many of these concentrated reagents are not to be taken from the hood after they have been dispensed. Thus open containers of concentrated acids or of ammonia (ammonium hydroxide) are not to be carried from the hood into the main laboratory area. If your laboratory has canopy hoods and you have to carry concentrated reagents to your station, place the quantity needed in a stoppered test tube and do not open the test tube until it is under cover of the canopy. Instructions in the manual will normally identify such steps as are to be carried out in the hood, but always work on the foregoing general principle.

Never pour concentrated acid or base directly from a reagent bottle into a reaction mixture or container holding any other chemicals, including water. A violent reaction may take place, causing you to drop the reagent bottle and possibly leading to serious injury. When a procedure calls for the addition of drops of concentrated acid or base to a reaction mixture the reagent will be stored in a dropper bottle or dispensing pipets will be provided.

Never directly smell any reagent; particularly do not inhale concentrated ammonia, organic solvents, or other volatile materials. When instructed to smell a gas or vapor, do so by gently wafting the air above the reagent vessel toward your nose and cautiously sniffing.

Less concentrated acids and bases are generally identified by having their molar or formal concentration marked on the container. Even dilute acids and bases can cause serious burns if they come in contact with skin or clothing. Handle such solutions with care and immediately flush any acid or base spilled on clothing or skin with copious amounts of cold running water. Use the safety showers in case of a massive spill on your person. Remove affected clothing promptly. Clean up all spills of acid or base on the bench top or floor using carbonate or bicarbonate powder to neutralize any acid spill and vinegar for any spilled base. Complete the clean up with copious amounts of cold water. Always inform your instructor of any large spill.

Never add water to acid. With concentrated acids a highly exothermic reaction is likely to occur, often with explosive violence. When adding acid to water, do so by cautiously adding a little at a time and swirling the container to ensure rapid mixing.

## Solid Reagents

Most of the reagents you will use in this course are of high purity and it is important that they remain in that condition. For this reason you should never insert a spatula or other implement directly into a reagent bottle. If the spatula is contaminated with some other chemical the entire bottle of reagent may become contaminated also. The proper procedure is to tip into the bottle cap the approximate amount of material you estimate will be required. Do not waste materials by over dispensing. Get into the habit of estimating quantities such as 0.5 g, 1 g, 5 g, etc. With a clean spatula, transfer material from the cap to your container and discard any excess left in the lid, disposing of it in accordance with instructions. Do not return possibly contaminated material to the reagent bottle. Replace the cap on the bottle and return the reagent to the proper place. Don't carry reagent bottles from the dispensing area to your bench but instead take a suitable container from your bench to the dispensing area. This ensures that your fellow students need not waste time searching the laboratory for a missing reagent.

## Liquid Reagents

Again our concern is to maintain the purity of the reagent in its container. Always pour from the reagent bottle into an intermediate clean, dry container. Never pour directly from the reagent bottle into a reaction mixture, unless from a dropper bottle using the dispensing dropper or dispensing pipet provided. If the bottle has a flat-topped glass stopper or a twist-off cap, always set the cap upside down on the bench top so that the underside does not become contaminated. Unlike solid reagents, liquids should never be poured into a bottle cap. If the bottle has a round stopper top, then keep hold of the stopper in one hand while pouring with the other. Always replace the stopper in the bottle. Do this even if others are waiting in line behind you to use the same bottle. Never leave the stopper for someone else to replace. Always return the reagent bottle to its proper place.

Remember that most organic liquids are quite volatile and that many of their vapors are flammable (i.e., they may catch fire and burn) and should never be exposed to open flames. In this experiment you will be working with dichloromethane and acetone. Be sure not to expose the vapors of either reagent to flames. If you must temporarily store either reagent on your bench top, do so either in a stoppered container or in a beaker covered by a watchglass. Should a fire develop in either of these containers, it may be extinguished by inverting a one-liter beaker over the flaming vessel so as to cut off the air supply. If a more serious fire develops, alert your instructor at once.

## The Importance of Safety, and a Reassurance

We have emphasized a number of hazards, how to avoid them, and how to deal with them if they arise. You may rest assured, however, that with proper attention to correct safety procedures and with care exercised at all times, all of the operations you are called on to perform in the laboratory may be carried out without danger to you or others. But the primary responsibility in observing safe practices rests with you. Always read the safety notes preceding each procedure and always pause to consider the safety aspects of any operation you are about to conduct. If there is the slightest doubt in your mind that what you propose to do may not be entirely safe, don't do it. Instead, consult your instructor whenever any question of safety arises in your mind. Above all, do not deviate from the written procedures in this or any other experiment unless directed to do so by your instructor.

**Remember to wear your safety goggles at all times in the laboratory.**

### 1. Preparation of the Sample and Preliminary Observations

Retrieve the sample in the sealed beaker into which you weighed it at the end of the last laboratory period. Examine the underside of the aluminum foil for signs of sublimed organic material. If any sublimate is present, note this fact on your data sheet and scrape it from the foil, returning it to the body of the sample or setting it aside in a sealed 30-mL beaker. In the latter case be sure to add this material to the organic sample once it has been recovered.

Note and record the general appearance of the sample, its color, crystallinity, and any obvious visible differences between the components, such as differences in color or particle size. Note any odor and the general consistency of the sample. Do *not* taste the sample.

## 2. Extraction of the Organic Component

Into a clean, dry 100-mL beaker pour 60–70 mL of dichloromethane. (*Work at the hood.*) Place a watchglass on top of the beaker to reduce evaporation. Add about 5 mL of the solvent to an 20-cm test tube, stopper the tube with a solid No. 5 rubber stopper, and shake it gently so as to rinse the inside of the tube and the stopper. Remove the stopper, discard the solvent in the labeled disposal container and repeat the rinse. There should be no yellowing of the solvent from contaminants in the rubber.

Transfer your sample of unknown from the beaker into the 20-cm test tube. Add 25 mL of dichloromethane and firmly stopper the tube. Keeping a firm grip on the stopper, invert the tube and its contents several times to mix the solvent with the solid. With the tube upright, open the stopper slightly to release any build up of pressure, replace the stopper, and repeat the shaking several times so as to promote solution of the organic component. Half fill a 250-mL beaker with hot water from the tap, loosen the stopper in the tube, and stand the tube upright in the beaker.

Now weigh on the rough balance the empty beaker and any mixture remaining in it. The difference between this weight and the weight recorded last week is the mass of sample transferred.

Set up a Buchner funnel and flask with a trap bottle in the line (see Figure 1.1). Be sure that the filter flask is secured in a stand and clamp and that the whole apparatus

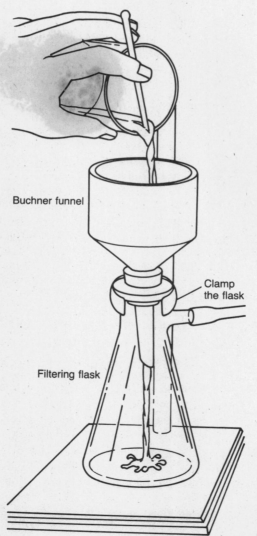

**Figure 1.1**
The use of a glass stirring rod in transferring a liquid solution to a Buchner funnel. The stirring rod should be touched against the inside wall of the funnel when the transfer is complete to collect the last drop.

is stable. Place a filter paper disk in the bed of the filter funnel and wet it with a little dichloromethane. Place your largest beaker under the aspirator outlet to minimize splashing, turn on the aspirator, and check that a good suction is obtained.

Remove the test tube containing the sample from the beaker, wipe off the outside of the tube, and decant off the supernatant liquid into the filter funnel down the side of a glass stirring rod, as shown in Figure 1.1. Add a further 25 mL of dichloromethane to the test tube and repeat the extraction process as before. Next, wet the filter paper with dichloromethane again and shake the test tube to mix the contents thoroughly and pour the mixture onto the filter bed under strong aspiration. Do not worry about any small amount of solid that may remain in the test tube; it may be retrieved and added to the main body of the inorganic residue when the solvent has completely evaporated from the tube. Be sure, however, that no solid passes through the filter into the filtrate. Continue aspiration until the residual solid is dry to the touch.

### 3. Recovery of the Organic Component

Place your initials on a clean, dry 100-mL beaker and weigh it to the nearest 10 mg. Disconnect the filter flask and funnel. Carefully remove the funnel and its contents and set them aside in a safe place. Transfer the filtrate from the flask to the beaker, rinsing the flask with a minimal amount of dichloromethane and transferring the rinsings to the beaker.

Take the beaker and its contents to the designated hood and store it there for the remainder of the period. If not too much dichloromethane has been used, all of the solvent should have evaporated by the end of the period, at which time you should remove the beaker. If time permits, the beaker and its contents should be weighed to the nearest 10 mg. Otherwise, cover the beaker top with a small piece of aluminum foil to minimize sublimation and store the beaker in your locker till the start of the next period. If evaporation is not complete, punch a few holes in the foil with a pen tip before storing the beaker away.

### 4. Extraction of the Water-soluble Inorganic Component; Isolation of the Insoluble Inorganic Component

Carefully remove the filter paper and its contents from the filter funnel and with a clean spatula transfer the solids to a clean 100-mL beaker. Add to the solids any dry solid remaining in the 20-cm test tube, using a glass stirring rod with a rubber policeman attached to effect the transfer. Be sure that all the solids are recovered.

Heat 25 mL of distilled water to boiling in a 50-mL beaker, using a tripod stand and wire gauze for support of the beaker. When the beaker has cooled sufficiently to allow it to be lifted by the rim, add the water to the solids in the 100-mL beaker (*caution*). Stir the mixture with a clean glass stirring rod for between three and four minutes and filter the mixture through the cleaned Buchner funnel in the same way as described in Step 2 above. Use distilled water from your wash bottle to wash any residual solids from the beaker into the filter funnel (see Figure 1.2).

Wash the solids in the filter bed with about 15 mL of hot distilled water and continue filtration until all the water has been drawn through the filter. Once again, no solids should pass through the filter paper.

Turn off the aspirator and set aside the funnel and its contents in a safe place. Transfer the filtrate to a clean 100-mL beaker. Reassemble the filter apparatus and wash the residual solid in the filter funnel with two 10-mL portions of acetone. Continue the aspiration until the odor of acetone is no longer noticeable above the solid. Discard the acetone filtrate in the labeled disposal container.

Weigh a clean, dry 50-mL beaker to the nearest 10 mg. With the aspirator turned off, carefully remove the filter paper and residual solid from the funnel, and with a clean spatula transfer as much of the solid as possible to the weighed beaker. Leave the beaker and its contents in your locker till next period so that they may dry completely, at which point the assembly should be reweighed to the nearest 10 mg to establish the mass of the insoluble inorganic component. At the same time, note its color and appearance.

## 5. Qualitative Analytical Tests

Tests for the presence of various common cations and anions are listed. You should carry out each test on the reference substances provided and then repeat it as necessary, substituting your unknown solution for the reference solution, so as to identify the constituent ions in each inorganic component. Record the results of each test on a reference substance on the data sheet and record any positive tests which result for your unknowns.

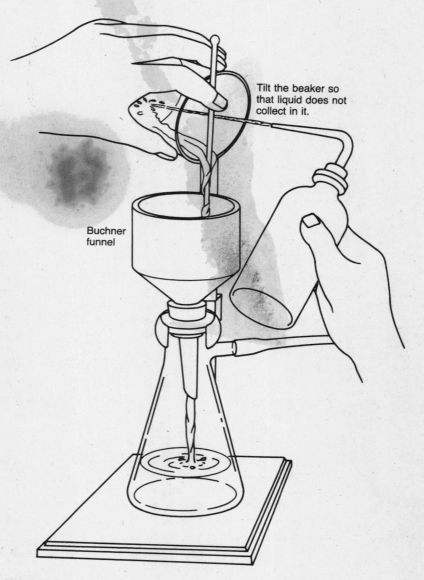

**Figure 1.2**
Using a washbottle to ensure complete transfer of a precipitate. Note that the Buchner flask should be secured in a stand and clamp.

### a. Cobalt(II), $Co^{2+}$

In aqueous solution the cobalt(II) ion exists as the hydrated species $[Co(H_2O)_6]^{2+}$, which imparts a pink color to the solution. Thiocyanate ion, $CNS^-$, will displace water from the hydrated $Co^{2+}$ to give the intensely blue-colored $[Co(CNS)_4]^{2-}$ complex ion:

$$[Co(H_2O)_6]^{2+}(aq) + 4CNS^-(aq) \longrightarrow [Co(CNS)_4]^{2-}(aq) + 6H_2O$$

Add 2 drops of $0.1M$ $Co(NO_3)_2$ to 1 mL of distilled water in a test tube. Add 5 drops of $1M$ $NH_4CNS$ and 1 mL of acetone and shake well. Note the result of the reaction on your data sheet.

### b. Copper(II), $Cu^{2+}$

Aqueous solutions of Cu(II) are light blue in color due to the presence of the hexahydrated $Cu^{2+}$ ion, $[Cu(H_2O)_6]^{2+}$. Addition of excess ammonia (ammonium hydroxide) displaces the loosely bound water molecules, replacing them with ammonia to yield the deep blue-colored cupric tetrammine complex ion $[Cu(NH_3)_4]^{2+}$:

$$[Cu(H_2O)_6]^{2+}(aq) + 4NH_3(aq) \longrightarrow [Cu(NH_3)_4]^{2+}(aq) + 6H_2O$$

To 5 drops of $0.1M$ $CuSO_4$ solution in a test tube add 5 drops of $4M$ ammonia solution. Shake well and record your observations.

### c. Iron(III), $Fe^{3+}$

The iron(III) ion in aqueous solution has a yellowish orange color. In acidic solution it reacts with thiocyanate ion to form a characteristic 1:1 complex ion, $Fe(CNS)^{2+}$, which has an intense red color diagnostic of Fe(III). The test is extremely sensitive. The red color is discharged by addition of fluoride ion, $F^-$:

$$Fe^{3+}(aq) + CNS^-(aq) \longrightarrow Fe(CNS)^{2+}(aq)$$

$$Fe(CNS)^{2+}(aq) + 6F^-(aq) \longrightarrow FeF_6^{3-}(aq) + CNS^-(aq)$$

Add one drop of $0.1M$ $Fe(NO_3)_3$ to 1 mL of $2M$ HCl in a test tube. To this mixture add 1 drop of $1M$ potassium thiocyanate. Note the result. To discharge the color add about 50 mg of solid sodium fluoride to the mixture and shake to dissolve.

### d. Nickel(II), $Ni^{2+}$

The hydrated nickel(II) ion gives rise to light-green aqueous solutions. The red complex formed by reaction between $Ni^{2+}$ and dimethylglyoxime in basic solution is used to identify this ion:

$$2 \begin{matrix} H_3C-C=N-OH \\ | \\ H_3C-C=N-OH \end{matrix} + Ni^{2+}(aq) + 2OH^-(aq) \longrightarrow \begin{matrix} O-H\cdots O \\ H_3C-C=N \diagdown \diagup N=C-CH_3 \\ | \quad\quad Ni \quad\quad | \\ H_3C-C=N \diagup \diagdown N=C-CH_3 \\ O\cdots H\cdots O \end{matrix} + 2H_2O$$

The formulas written for dimethylglyoxime and its nickel complex are known as structural formulas and make use of lines to indicate chemical bonds. You will learn about structural formulas in Chapter 18 of *Chemical Principles*.

Add 1 drop of $0.1M$ $NiCl_2$ to 1 mL of $2M$ aqueous ammonia solution in a test tube and shake well. Add 5 drops of 1% dimethylglyoxime solution, shake well, and record the result.

### e. Ammonium, $NH_4^+$

The action of strong bases on aqueous solutions of ammonium salts results in formation of gaseous ammonia detectable by its odor:

$$NH_4^+(aq) + OH^-(aq) \longrightarrow NH_3(aq) + H_2O$$

Add 1 mL of 4$M$ NaOH to 1 mL of 1$M$ NH$_4$Cl in a test tube, shaking to mix well. Bring the contents to a gentle boil and very carefully, with your hand, waft the gas produced toward your nose and sniff gently.

Some people cannot smell ammonia. If you are one of them, adopt the following alternative test. Take a 3-cm length of pH paper and moisten it with a drop of distilled water. Hold the moist paper above the mouth of the test tube during the heating step. The development of an intense blue color on the paper indicates the presence of ammonia gas.

### f. The Alkali Metal Ions, Na$^+$ and K$^+$, and the Alkaline Earth Ion Ca$^{2+}$

These ions are detected by the characteristic colors in a flame test. Sodium produces an intense yellow-orange color, potassium a blue-violet color, and calcium a bright brick-red color. Confirm these colors for the reference substances as follows.

Dispense about 3 mL of concentrated nitric acid (*at the hood*) into a clean, dry test tube, stopper the test tube, and carry it back to your bench where you should stand it upright in a 250-mL beaker. Take a 15-cm length of nichrome wire and at one end make a small loop about 1 mm in diameter. Dip the loop into the concentrated nitric acid in the test tube, withdraw it, and place it in the hottest part of a clear blue burner flame (about 1 cm above the tip of the central cone), leaving it in the flame until all contaminant colors have burned out and the only color present is the red hot color of the wire. Repeat this purification of the wire as many times as necessary. Place the wire where it will not become recontaminated.

Take a very small amount of the reference solid on the tip of a clean spatula, dip the wire into the concentrated HNO$_3$, and on its tip pick up a small quantity of solid. Place the wire tip in the flame as before and observe the color produced, first with the naked eye and then through a piece of blue cobalt glass. Because the sodium color is particularly persistent, it is best to test in the sequence: potassium, calcium, sodium.

For your unknown carbonate, repeat the test as described above, substituting a very small amount of the previously weighed unknown. For the soluble inorganic, use a drop of the solution but omit the final dip of the wire into the nitric acid. In this case the colors produced may not be as intense as for the references.

Dispose of the residual concentrated HNO$_3$ by pouring it carefully down one of the sink drains in the hood and flushing the drain with copious amounts of cold running water.

### g. Halide Ions, Cl$^-$, Br$^-$, I$^-$

In acidic aqueous solution all of these ions yield insoluble precipitates with Ag$^+$ and this test is used as a general diagnostic for the presence of halide ions:

$$Ag^+(aq) + X^-(aq) \longrightarrow AgX(s),$$

where X$^-$ represents the halide.

Some discrimination may be achieved by noting that AgCl is white, AgBr a cream color, and AgI yellow, but this observation is not conclusive, particularly where more than one halide is present. For a more reliable identification of the individual halides, use is made of their differential solubility in 2$M$ ammonia solution with which they form ammine complexes:

$$AgCl(aq) + 2NH_3(aq) \longrightarrow [Ag(NH_3)_2]^+(aq) + Cl^-(aq)$$

Carry out the following tests on the three reference solutions and note the results. Use 0.1$M$ NaCl, KBr, and KI as references.

Label three glass centrifuge tubes to identify the three references. To each add 5 mL of distilled water and 5 drops of the reference solution, followed by 10 drops of 3$M$ nitric acid (*not* concentrated nitric acid) and 10 drops of 0.1$M$ AgNO$_3$. Arrange the three tubes symmetrically in the centrifuge, or balance them with tubes containing an equal volume of distilled water, and centrifuge the mixtures for about 5 minutes. Note the colors of the precipitates. Discard the supernatant liquid and add 2 mL of 2M ammonia solution to the solid residue in each tube. Shake each well and note the result. Add 3 mL of 3M nitric acid to each tube, shake well and again record the result.

### h. Carbonate Anion, $CO_3^{2-}$

All carbonates react with strong acids to liberate carbon dioxide as a gas. $CO_2$ reacts with lime water, a solution of $Ca(OH)_2$, to form insoluble calcium carbonate:

$$CO_3^{2-}(aq) + 2H^+(aq) \longrightarrow CO_2(g) + 2H_2O$$

$$H_2O + CO_2(g) + Ca^{2+}(aq) \longrightarrow CaCO_3(s) + 2H^+(aq)$$

Disconnect the longer of the two right-angled glass tubes in the trap bottle and insert the shorter end into a one-hole stopper that fits the test tube (No. 0). No more than 3 cm of the glass tube should protrude into the test tube. Add 3 mL of $0.1M$ $Na_2CO_3$ to the test tube and support the test tube with a clamp and stand, as shown in Figure 1.3. Half fill another test tube with lime water. To the test tube containing the carbonate carefully add a few drops of 6M HCl. Stopper the test tube with the glass tube assembly and insert the other end of the tube into the test tube containing the lime water so that the end of the glass tube reaches just below the surface of the lime water. Gently heat the carbonate solution to drive off $CO_2$, which will bubble into the lime water, turning it cloudy.

### i. Nitrate Ion, $NO_3^-$

When nitrates are treated with strongly acidified diphenylamine, a deep blue-violet complex is formed.

Add 5 drops of sulfuric acid–diphenylamine solution to an equal quantity of $0.1M$ $KNO_3$ in a test tube, shaking to mix. Carefully (hood) add 10 drops of concentrated sulfuric acid to the mixture in the test tube and swirl very gently to mix the reagents.

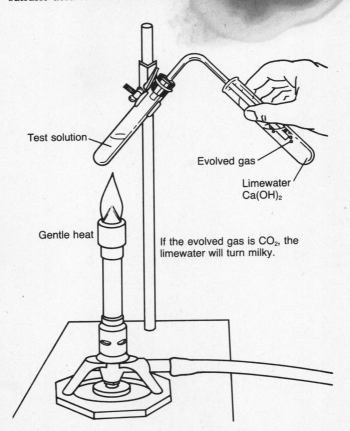

**Figure 1.3**
Testing for $CO_2$ with limewater. The end of the glass tube in the test tube containing the lime water should be just below the surface and the test solution should be heated gently.

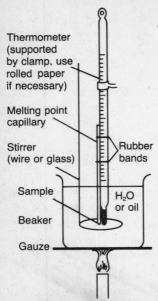

**Figure 1.4**
A melting point apparatus.

### 5. Sulfate Ion, $SO_4^{2-}$

Sulfate ion is recognized by the formation of insoluble barium sulfate, when solutions containing these two ions are mixed:

$$Ba^{2+}(aq) + SO_4^{2-}(aq) \longrightarrow BaSO_4(s)$$

Add 1 mL of $3M$ HCl (*not* concentrated HCl) to 1 mL of $0.1M$ $Na_2SO_4$ solution in a test tube. Shake well and warm the solution for a minute or so in a gentle burner flame, then add 10 drops of $0.1M$ $BaCl_2$ solution and shake well.

### 6. Melting Point Determination and Identification of the Organic Component

Prepare a melting point tube by sealing one end of a capillary tube in a hot flame. Allow the tube to cool.

Take a small quantity of your previously weighed organic component and with a spatula or glass stirring rod grind it to a fine powder on a watchglass. Pack enough of the finely divided solid into the melting point tube to fill the bottom half inch or so. Tap the tube to coax the solid to the bottom.

Fill a 250-mL beaker about two-thirds full of water and place it on a wire gauze on a ring stand above a burner. Place the thermometer in a one-hole rubber stopper after lubricating the stopper hole with a drop of glycerin. With a small rubber band attach the melting point tube to the stem of the thermometer so that the tube is parallel to the stem and the bottom is directly opposite the bulb. Secure the thermometer to the same stand with a clamp and adjust its height so that the bulb is centrally located in the water.

Stir constantly as you heat the beaker, observing the behavior of the solid in the tube. Record the temperature at which the first sign of melting appears and that at which melting is complete. The process may be repeated if necessary by allowing the sample to cool and resolidify. Identify your unknown from the posted list of melting points.

Prepare an intimate mixture of equal amounts of your unknown and the reference compound you believe it to be and carry out a mixed melting point determination in the same way as before. Remember to keep your spatula out of the reagent bottle containing the pure reference compound and take only as much of the reference compound as you will actually need.

If the melting point of your mixture is the same as that of your unknown (within about 0.5 K) you may assume that you have correctly identified the organic component. If the melting point of the mixture is significantly lower than that of the unknown, then the identification is probably wrong or your organic fraction is contaminated. Check the original melting point of the unknown and your identification. If there is another possible reference compound with a melting point close to that of your unknown, repeat the mixed melting point test with that reference. Be sure to record *all* melting point determinations on your data sheet.

### 7. Estimating and Testing the Insoluble Carbonate

The mass of the insoluble carbonate should have been determined in Step 4. If you have not done this do so now following the directions given there.

Only the cation present in the carbonate need be identified. Take a few milligrams of the solid on the tip of your spatula and dissolve it in 1 mL of $3M$ nitric acid in a test tube. It may be necessary to heat the test tube gently to ensure complete solution. One drop of this solution may be used for each cation test, but in cases where the test requires an alkaline solution (e.g., for $Ni^{2+}$, $Cu^{2+}$) it may be necessary to increase the quantity of base specified in the test so as to ensure that the acidic carbonate solution has been fully neutralized. Use the information on colors and solubility given earlier to guide your choice of tests. Only one cation should be present.

## 8. Estimating the Soluble Inorganic Component

The mass of the soluble inorganic component is found by subtracting the combined masses of the isolated organic and insoluble inorganic components from the original mass of sample taken. Although you could evaporate the solution containing the soluble inorganic component to dryness and weigh it, that is quite difficult to do quantitatively, as you will see in Experiment 4. You might also recrystallize the salt in a hydrated form different from that with which you started, which could give misleading results. However, if you wish, you may set aside the solution containing the soluble inorganic salt at the back of your locker and observe next week the solid that forms on slow evaporation.

## 9. Cleanup and Waste Disposal

By the end of each laboratory period you should clean or wash and dry all equipment used in the period. Equipment checked out of the stockroom should also be cleaned and returned, preferably as soon as you have finished using it, but in any event before the stockroom closes.

You should wash down and dry your bench top with the sponges provided.

The organic solvents used in this experiment must be disposed of safely. Special labeled containers are provided in the hoods for any leftover dichloromethane and acetone. Take care not to get the solvents mixed up when disposing of them.

At the concentrations used here none of the inorganic salt solutions are harmful if washed down the drains with copious amounts of water. Inorganic solids should be placed in the labeled containers or in special waste buckets, following the instructor's directions.

5. What precautions should you take in handling bottles of liquid reagents?

6. In testing for cation type in the insoluble carbonate you are not expected to test for $Na^+$, $K^+$, or $NH_4^+$. Why is that?

7. Why do you suppose that the sample must be close to the thermometer bulb in determining the melting point of your organic unknown?

8. What is meant by a mixed melting point and how is it diagnostic of purity?

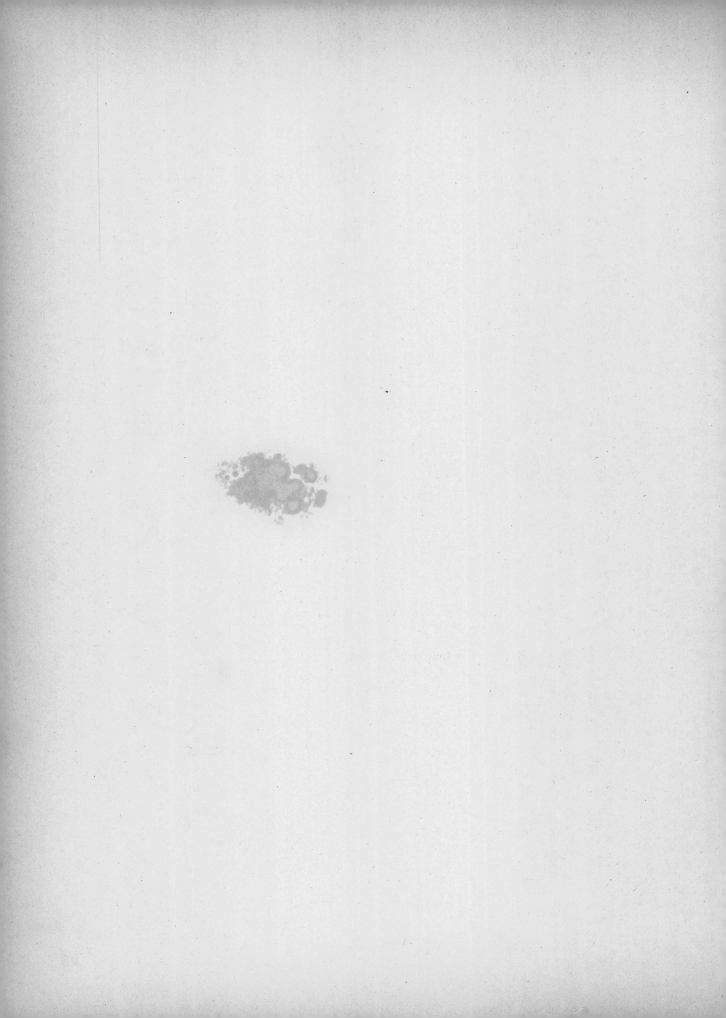

Name _____     Instructor _____

# DATA SHEET

Use carbon paper to make duplicate entries. Do not fill in the two copies of the data sheet separately. Be sure to provide all units where required, e.g., g, °C, etc.

1. Enter the mass of the beaker + sample from the data sheet for last week's experiment and the mass of the beaker alone.

|  | Mass | Possible error |
|---|---|---|
| Beaker + sample | _____ ± | _____ |
| Beaker alone | _____ ± | _____ |
| Sample | _____ ± | _____ |

2. Enter your observations on the sample as presented.

   *blue liquid, no smell,* _____

   _____

3. Organic component:

|  | Mass | Possible error |
|---|---|---|
| Beaker + dry solid | _____ ± | _____ |
| Beaker alone | _____ ± | _____ |
| Dry solid | _____ ± | _____ |

Color and appearance of solid:

_____

Observed melting point of solid:

Run 1 _____ to _____ ± _____

Run 2 _____ to _____ ± _____

Inferred identity: _____

Melting point of reference: _____

Mixed melting point: _____ to _____ ± _____

Enter any additional measurements needed:

_____

_____

Name _____ Instructor _____

## DATA SHEET

Use carbon paper to make duplicate entries. Do not fill in the two copies of the data sheet separately. Be sure to provide all units where required, e.g., g, °C, etc.

1. Enter the mass of the beaker + sample from the data sheet for last week's experiment and the mass of the beaker alone.

|  | Mass |  | Possible error |
|---|---|---|---|
| Beaker + sample | _____ | ± | _____ |
| Beaker alone | _____ | ± | _____ |
| Sample | _____ | ± | _____ |

2. Enter your observations on the sample as presented.

   _____

   _____

3. Organic component:

|  | Mass |  | Possible error |
|---|---|---|---|
| Beaker + dry solid | _____ | ± | _____ |
| Beaker alone | _____ | ± | _____ |
| Dry solid | _____ | ± | _____ |

   Color and appearance of solid:

   _____

   Observed melting point of solid:

   Run 1  _____ to _____ ± _____

   Run 2  _____ to _____ ± _____

   Inferred identity: _____

   Melting point of reference: _____

   Mixed melting point: _____ to _____ ± _____

   Enter any additional measurements needed:

   _____

   _____

4. Note your observations on the qualitative analytical tests carried out on the reference reagents.

**Cations**

$Co^{2+}$ _____

$Cu^{2+}$ _____

$Fe^{3+}$ _____

$Ni^{2+}$ _____

$NH_4^+$ _____

$Na^+$ _____

$K^+$ _____

$Ca^{2+}$ _____

**Anions**

$Cl^-$ _____

$Br^-$ _____

$I^-$ _____

$NO_3^-$ _____

$CO_3^{2-}$ _____

$SO_4^{2-}$ _____

5. Soluble Inorganic Component:

   Color of solution: _clear white_____

   Place a circle around each ion tested for and describe the outcome of any tests yielding positive results. Only one anion is present but there may be more than one cation. Do no more tests than you deem necessary.

   ⓒo²⁺  ⒸCu²⁺  ⒻFe³⁺  ⓃNi²⁺  ⓃH₄⁺  ⓃNa⁺  ⓀK⁺  ⒸCa²⁺
   ⒸCl⁻   Br⁻   I⁻   NO₃⁻   CO₃²⁻   SO₄²⁻

   _____

   _____

6. Insoluble carbonate component:

   |  | Mass | Possible error |
   |---|---|---|
   | Beaker + dry solid | 25.12 ± | 0.01 |
   | Beaker alone | 19.94 ± | 0.01 |
   | Dry solid | 5.18 ± | 0.01 |

Name _____    Instructor _____

## REPORT SHEET

Be sure to include all relevant units of measurement. Check all calculations carefully. Use the proper number of significant figures in derived results as well as in measurements.

*del.* 1. Enter from your data sheet:

|  | Mass | | Possible error |
|---|---|---|---|
| Total sample | _____ | ± | _____ |
| Organic component | _____ | ± | _____ |
| Insoluble inorganic | _____ | ± | _____ |

2. Calculate by difference the mass of the soluble inorganic component:

_____ ± _____

3. Express the masses of the three components as percentages of the mass of the original sample:

    **a.** Organic component _____ ± _____ %

    **b.** Insoluble $CO_3^{2-}$ _____ ± _____ %

    **c.** Soluble inorganic _____ ± _____ %

4. Give the identity of the organic component.

_____

5. Give the name and formula of the soluble inorganic salt.

_____

6. Give the name and formula of the insoluble carbonate.

_____

7. The errors in the percentages given in Item 3 above are based only on uncertainties in measurement. Describe any other sources of error inherent in the methods used and assess their contribution in absolute and percentage terms to the results obtained.

_____

_____

_____

_____

_____

## WORKSHEET

Carry out all the calculations for Experiment 1 on this sheet, attach it to your report sheet, and submit the two together. Do your calculations directly on this sheet. Do not do them on scratch paper and then copy them on to this sheet. Make no erasures and delete incorrect entries with a single line so as to leave the original data legible. Use a pen.

# EXPERIMENT 2

# Chromatographic Separation and Identification of Metal Ions in Aqueous Solution

### AIM

To demonstrate the selective adsorption of metal ions in paper chromatography and by selective adsorption to identify the metal ions in an unknown mixture.

### EQUIPMENT

From locker:  50-, 250-, and 1000-mL beakers
25-mL graduated cylinder
No. 5 solid rubber stopper
Glass stirring rod
Ruler
Watchglass

### REAGENTS AND OTHER MATERIALS

$0.1M$ acidified solutions of Co(II), Cu(II), Fe(III), Mn(II) & Ni(II) salts with dispensing pipets

**EXPERIMENT 2**

Concentrated HCl and acetone as mobile phase
Concentrated $NH_3$, $H_2S$, and developing sprays
Whatman No. 1 filter paper

## BACKGROUND

### Associated Readings

The following sections of *Chemical Principles* contain material useful in understanding Experiment 2. Read them before coming to the laboratory and again after you have completed the experiment, to get maximum benefit from your laboratory work.

**Chapter 1** The Science of Chemistry
            **1.2** Chemistry as a Physical Science
**Chapter 2** The Language of Chemistry
            **2.4** Balanced Chemical Equations
            **2.5** Important Types of Chemical Reactions
            **2.6** States of Substances

In Experiment 1 you used differences in their solubility in two different solvents to separate the components of a heterogeneous mixture. Frequently, however, one is confronted by homogeneous mixtures in which all the components are soluble in the same solvent, as are many metal salts in water. To effect a separation in such a case, you must make use of some other physical property by which the components may be distinguished. A particularly useful property is the different extent to which different substances are **adsorbed** on the surface of a given material.

Differential adsorption, as this property is known, was first exploited in chemistry as a purification tool. Solutions of technical-grade colorless compounds are often colored, the color arising from various byproducts of the original synthesis present as impurities. If finely divided charcoal is added and the solution is refluxed and then filtered, the resulting filtrate is usually colorless indicating that the colored impurities have been trapped or adsorbed upon the surface of the charcoal. One should distinguish, at this point, between **adsorption**—the attachment of chemical species to a surface—and **absorption**—a process in which chemical species infiltrate the body of a material, as when a hygroscopic salt or liquid takes up water from the air.

The analytical method that exploits differences in adsorption is known as **chromatography**. This use of the word has completely supplanted the original meaning—the description of colors—but it recalls the earliest use of the method by the Russian scientist Tsvett, in 1903, to separate the different colored pigments in plant extracts. There are many different forms of chromatography in use today, making possible, for example, the separation of the components of a mixture of gases or volatile liquids (**gas chromatography**) or the separation of different solids dissolved in the same solvent (**liquid chromatography**). It is the most widely used method for separation of a complex homogeneous mixture. Modern chromatographic methods are highly sensitive, allowing the separation of amounts of material in the microgram range. All variants of the method share certain common features.

First an adsorbent surface, a **stationary phase,** must be provided which, as its name implies, remains in a fixed position during the separation. The most useful stationary phases all have a large surface area in relation to their volume but, that aside, the best choice of stationary phase for a particular problem is often a matter of trial and error. Several materials—such as activated charcoal or silica gel—are widely used in the separation of gases. One gram of finely divided charcoal, activated by prolonged heating at around 600°C in an inert atmosphere, can adsorb about 350 mL of $SO_2$ at room temperature, for example. For liquid systems, alumina powder (finely divided $Al_2O_3$), first used by Tsvett, or cellulose (filter paper) are frequently used.

The second stage in the process involves the separation of the components of the mixture by passage of the **mobile phase** in which they are carried through or across the stationary phase. In liquid chromatography a small amount of the mixture is deposited at some particular location on the stationary phase, such as at the top of a column of alumina or the bottom of a piece of filter paper. The molecules (or atoms or ions) of the substances adsorbed are held to the surface molecules of the stationary phase by a variety of intermolecular forces of different strengths. If the various components are now exposed to a uniform flow of some solvent, they will be washed along the surface of the adsorbent but at different rates, which reflect the degree of attachment of their molecules to those of the adsorbent versus their affinity for the solvent. Less tightly bound molecules will move more rapidly, and hence move farther in a given time, than will more tightly bound ones. As all of the molecules of a given species will be bound to the same degree to the surface, they will tend to cluster together and so a separation of the different components will be achieved.

In **column chromatography,** where the mixture is placed at the top of the column and the various components washed down at different rates, the process may be continued until each component is washed out of the column in turn and collected. Alternatively, the pieces of the column to which the different components are bound may be physically separated and a different solvent used to wash the component free. This process is known as **elution.**

In **paper chromatography,** the method we will use, the mixture is adsorbed on a piece of filter paper and the mobile solvent front is allowed to ascend the strip by capillary attraction, carrying the various components with it until the requisite separation has been achieved. The paper may then be dried and the parts to which the components are adsorbed cut out and eluted to remove the pure material.

In cases where only identification is required, rather than retrieval of the individual components, the solvent flow can be halted when an adequate separation has been achieved but before the solvent front reaches the extremity of the stationary phase. The ratio of the distance traveled by a particular component to that traveled by the solvent front is known as the **retardation factor,** $R_f$, and is characteristic of the particular component and of the stationary and mobile phases used.

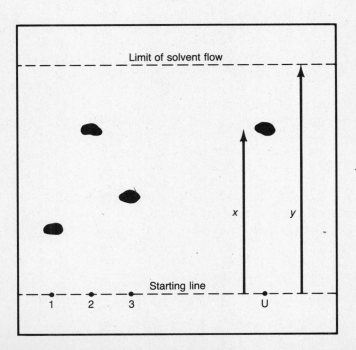

**Figure 2.1**
Developed paper chromatogram showing three reference spots and, on the right, an unknown. The $R_f$ value for the unknown ($x/y$) matches that for the middle reference substance.

Recognition of how far the various components have traveled is straightforward when they are colored substances but is more difficult when they are colorless or the same color as the stationary phase. In such a case an extra step is needed—**development**. Colorless components may be located by either physical or chemical means. For example, if the material fluoresces it may be seen by shining an ultraviolet lamp on the stationary phase. Alternatively, the stationary phase may be treated chemically with some reagent, which yields a colored product with the components. The reagent ninhydrin, for example, reacts with peptides to give intensely purple products, and is widely used in protein analysis.

Although these basic features are common to all forms of chromatography, the $R_f$ identification method, so called because the stationary phase is a piece of ordinary filter paper (cellulose), is most widely used in paper chromatography and in a related technique called **thin-layer chromatography** in which the adsorbent is a thin film of alumina or silica gel spread over a glass support. Paper chromatography is cheap and highly versatile. It was used in prize-winning work by many Nobel laureates: by Frederick Sanger in his determination of the amino acid sequence in the hormone insulin; by Melvin Calvin, who combined the technique with the use of radioactive tracers in his elucidation of the mechanism of photosynthesis in plants; and by the English biochemists A. J. P. Martin and R. L. M. Synge who were awarded the 1952 prize in chemistry for their work in developing and exploiting the technique.

In the example shown in Figure 2.1, the solution to be analyzed is known to contain one of three possible substances. A drop of the solution is placed on the right-hand side of the filter paper and a drop of solution of each of the three possible pure substances is placed at the same height on the left-hand side of the paper and labeled. The paper is placed in a dish of suitable solvent, as shown, and capillary action draws the solvent up the paper and with it the three reference substances—at different rates—and the unknown. The unknown will have the same $R_f$ value as one of the references because, being chemically identical, they will have traveled at the same rate.

This scheme is delightfully simple, but a variety of highly ingenious chromatography procedures have been developed and combined with other separation methods such as **electrophoresis**—a technique whereby adsorbed ionic substances are separated by means of an electrical current acting on the ionic charges. Protein analyses are often carried out by chromatography in one direction followed by electrophoresis in a direction at right angles to the first so that a two-dimensional resolution of the peptide fragments produced on digestion of the protein is obtained. Exciting new developments in this area are now taking place in space. In the space shuttle *Challenger* experiments involving zero-gravity chromatography and electrophoresis have shown that complex mixtures of medicinally active natural products can be separated with much greater ease and efficiency than on earth.

Back on earth, in this laboratory period we use a simple paper chromatography technique to separate and identify a mixture of inorganic cations in aqueous solution. Despite its simplicity, the experiment embodies all of the principles described earlier. As a stationary phase we use a filter paper support. The mobile phase is a mixture of acetone and hydrochloric acid. Depending on the cations assigned, the chromatogram is developed in one of three ways: with ammonia and $H_2S$ to give colored sulfides, with a developing spray that gives colored complexes, or with a spray that yields spots visible under ultraviolet light.

## PROCEDURE

### SAFETY NOTES
Handle concentrated hydrochloric acid and ammonia only in the hood and with great care. Neutralize any acid spills with the carbonate or bicarbonate powder provided. Rinse any acid spilled on skin or clothing with copious amounts of cold running water.

# Chromatographic Separation and Identification of Metal Ions in Aqueous Solution

**Acetone is a highly volatile flammable liquid. Keep open flames away from this reagent.**

Dispose of any leftover acetone-hydrochloric acid mixture in the labeled waste container in the hood. Do not dump this mixture in the sink or drains.

At the hood add 1 mL of concentrated hydrochloric acid from the dispenser to 19 mL of acetone in a 25-mL graduated cylinder. Stir the resultant mixture with a glass stirring rod to homogenize the contents. Stopper the graduated cylinder and take it to your bench.

Cut out three 8 × 8 cm squares of chromatographic paper (Whatman No. 1 filter paper). On each rule a light pencil line parallel to and 1.5 cm from one edge. On one sheet mark a series of evenly spaced pencil dots equal in number to the number of reference cations in the set assigned you by the instructor. The dots should be at least 1 cm apart and should be just below the pencil line. Avoid placing a dot at the center of the line. Under each dot, pencil in the chemical symbol or some other identification for one member of the set of reference substances. Now fold the paper evenly in two along an axis at right angles to the pencil line so that the two halves make an angle of about 60°.

Using the dispensing pipets provided with each reagent, place a very small drop of the appropriate reference reagent on the line just above the labeled dot. Do this by inserting the pipet into the reference solution, then withdrawing it, so that the tip is just above the solution. Allow the solution to drain from the pipet back into the bottle and when no further drop falls use a piece of tissue paper to remove all but about 1.5 mm of solution from the tip. Apply the tip of the pipet to the paper. Too large a drop will not give good results.

Add 5 mL of the solvent mixture (acetone-HCl) to a clean, dry 250-mL beaker. Check that the spots on the paper are not quite dry and stand the folded paper upright in the beaker with the line of spots at the bottom. The solvent level should not be above the pencil line, else the materials will leach directly into the bulk of the solvent; and the paper should not touch the sides of the beaker or solvent flow will be uneven. Cover the beaker with a watchglass so as to reduce evaporation of the solvent and allow the solvent front to rise to about 1 cm from the top of the paper.

Remove the paper with your fingers, holding it by the dry top, and with a pencil mark the limit of travel of the solvent front. Let the paper dry and circle any visible spots. Further develop the chromatogram in the manner recommended by your instructor. Circle the resulting spots and record their colors on your data sheet.

Measure and record the distance traveled by each spot from the starting line as well as the distance traveled by the solvent front. Use this information to derive the $R_f$ value of each reference.

Repeat the procedure on a fresh square of paper, using a drop of your unknown at each of three locations on the starting line. Develop the resulting spots and determine their average $R_f$ values. From these values and the observed colors of the spots, identify the unknown cations present in the mixture.

Check your identifications on a third chromatogram, this time placing a drop of your unknown on the left-hand side and labeled drops of the presumed references on the right. After development and drying, the chromatogram should be attached to your completed data and report sheet. Submit the two to your instructor before you leave the laboratory.

As a final exercise, your instructor may assign you an unknown mixture yielding only colored spots. Carry out a separation in the same way as before, but instead of developing the spots, cut them out and drop them each into separate test tubes containing 2 mL of distilled water. Shake the test tube to aid in leaching out the salt into the water and use the solution to identify the cation by means of the tests given in Experiment 1.

Name _____  Instructor _____

## PRELABORATORY PREPARATION SHEET

1. What is the function of the *stationary phase* in chromatography?

   _____
   _____

2. What is meant by the term *elution* in chromatography?

   _____
   _____

3. What is meant by the term *development* in chromatography?

   _____
   _____

4. What is the meaning of the term *retardation factor* in chromatography?

   _____
   _____

5. In an experiment in paper chromatography, a spot of an unknown traveled 5.0 cm, whereas the solvent front traveled 12.0 cm. With the same stationary phase and solvent system, two references A and B traveled 3.2 cm and 4.6 cm, whereas the solvent front traveled 11.0 cm. To which of the two references does the unknown correspond? Show your reasoning.

   _____
   _____
   _____

## WORKSHEET

Carry out all the calculations for this Experiment 2 on this sheet, attach it to your report sheet, and submit the two together. Do your calculations directly on this sheet. Do not do them on scratch paper and then copy them on to this sheet. Make no erasures and delete incorrect entries with a single line so as to leave the original data legible. Use a pen.

# EXPERIMENT 3

# Determination of the Empirical Formula of a Metal Sulfide

### AIM

To establish the empirical formula of a metal sulfide by determination of the mass of sulfide formed by combination of sulfur with a known mass of metal

### EQUIPMENT

| | |
|---|---|
| From stockroom: | Two porcelain crucibles with lids |
| From locker: | Two clay triangles |
| | Crucible tongs |
| Laboratory: | Bunsen or Fischer burner |
| | Tripod or ring stand |
| | Wooden splints |

### REAGENTS

Metal granules, wire, or chips
Powdered sulfur,
Ferric chloride solution for marking crucibles

## BACKGROUND

### Associated Readings

The following sections of *Chemical Principles* contain material useful in understanding Experiment 3. Read them before coming to the laboratory and again after you have completed the experiment to get maximum benefit from your laboratory work.

**Chapter 3** Stoichiometry
    **3.1** The Mole and Avogadro's Number
    **3.2** Mass Relationships in Chemical Formulas

---

The chemical atomic weight scale expresses the relative masses of the atoms of each of the elements. The masses are expressed relative to the mass of a single atom of the $^{12}C$ isotope of carbon, which is assigned a mass of exactly 12 unified atomic mass units (12 u). Thus, one unified atomic mass unit (1 u) has a mass exactly one-twelfth of the mass of a single atom of $^{12}C$. An atom having a mass of 36 u has three times the mass of an atom of $^{12}C$.

Whereas the unified atomic mass unit is suitable for single atoms, it is not immediately useful for describing macroscopic quantities of substances. Chemists measure quantity in terms of mass, most often in the gram range. Gram quantities of an element contain very large numbers of atoms of the element and we need a unit with which we can define the amount of an element present in terms of the number of atoms it contains. The unit used is the **mole.** A mole of an element is the amount of the element that contains as many atoms as there are atoms of $^{12}C$ in exactly 12 g of $^{12}C$. That number of atoms, known as **Avogadro's number,** may be found with considerable accuracy from X-ray interferometric measurements, but is approximated for most practical purposes to $6.0221 \times 10^{23}$.

The mass, in grams, of 1 mole of $^{12}C$ atoms is known as the molar mass of $^{12}C$. Because both the mole and the unified mass unit are defined relative to $^{12}C$, the relationship between the mass of an atom of $^{12}C$ and a mole of $^{12}C$ also holds for the other elements. The molar mass of any other element is the mass, in grams, of one mole of its atoms (with proper allowance for natural isotopic distribution). Correspondingly, the mass in grams of one mole of an element is numerically equal to the mean mass of its atoms in unified atomic mass units.

We may use the relationship

$$\frac{\text{mass in grams of element}}{\text{molar mass of element}}$$

to find the amount, in moles, of an element present in a given sample.

The formula weight of a compound is simply the sum of the atomic weights of the atoms present in the compound, as indicated by the chemical formula, and is numerically equal to the molar mass of the compound—the mass of one mole of the compound expressed in grams. For a molecular compound, the formula weight corresponds to the molecular weight. For an ionic compound it corresponds to the sum of the atomic weights of the elements present in the simplest chemical formulation of the substance. For calcium chloride, $CaCl_2$, for example, the formula weight is equal to the atomic weight of calcium plus two times the atomic weight of chlorine, i.e., $40.08 + 2 \times 35.45 = 110.98$, and is numerically equal to the mass in grams of one mole of calcium chloride.

For a metal sulfide produced by the reaction

$$x\text{M} + y\text{S} \rightarrow \text{M}_x\text{S}_y,$$

one mole of the sulfide contains $x$ moles of atoms of the metal M and $y$ moles of atoms of sulfur (in the sulfide present as $S^{2-}$ ions). If the masses of M and S present in a sample of the sulfide are known, then the ratio $x:y$ may be readily found from the equality

$$\frac{\text{moles M}}{\text{moles S}} = \frac{\text{number of atoms M}}{\text{number of atoms S}} = \frac{x}{y}$$

The amount (in moles) of M, of course, is simply the mass of M divided by its molar mass. The amount of S, similarly, is the mass of sulfur divided by 32.06 g mol$^{-1}$, its molar mass.

It is important to recognize that only the ratio $x:y$ is determined in this way and not the actual individual values of $x$ and $y$. For ionic compounds whose formulas are expressed in any case in terms of the simplest relative whole-number ratios of the amounts of the various ionic species present (e.g., $BaI_2$ rather than $Ba_2I_4$), the method normally gives a correct formulation in conventional terms. However, for molecular compounds the reasoning outlined above would not allow a distinction between, say, $C_3H_{12}O_3$ and $C_6H_{24}O_6$—two different compounds—but would yield only the formulation $CH_4O$. The formula of a molecular compound expressed in terms of the simplest whole-number ratio between the numbers of atoms of the different elements present in the molecule is known as its **empirical formula**.

It should be noted that much of the usefulness of the method described for determining the empirical formula of a compound arises because the molar ratios are, in general, the ratios of *small* whole numbers. This is just as well because our experimental measurements frequently contain errors of a few percent so that rounding off may be required to obtain the true whole-number ratio. In arriving at our ratios, however, we must exercise care not to round off beyond the likely limits of error. Thus a ratio of 1.99:1.00 with a 2% error might legitimately be rounded off to 2:1, but a ratio of 1.86:1.00, subject to the same error, should be interpreted as indicating a true ratio of 9:5 rather than 2:1. While small whole-number ratios are the rule for ionic compounds, less simple ratios are sometimes found and in extreme cases the ratios are not at all simple. An observation of Cu:S = 1.97:1.00 with an appropriately small uncertainty might indeed require interpretation as $Cu_{63}S_{32}$, the empirical formula of the mineral djurleite.

Extremely painstaking and accurate work is required to establish such complex ratios and the experimental uncertainties in this experiment do not justify considering such possibilities. However, it is important not to bend experimental observations to fit a preconception of undue simplicity. At all times exercise a scrupulous intellectual honesty in interpreting your experimental results and take into account the likely magnitude of your experimental errors and how they may have affected your answers. Such habits will stand you in good stead when you are confronted by new or unknown compounds.

## Sample Calculation

The following worked example shows how the results of an experiment using this method might be interpreted.

> A 0.977 g sample of calcium metal was reacted in a sealed system with excess chlorine gas and completely converted to 2.695 g of calcium chloride. The atomic weights of Ca and Cl are 40.08 and 35.45, respectively. Calculate the empirical formula of calcium chloride.

Mass of calcium chloride = 2.695 g
Mass of calcium = 0.977 g
Mass of chlorine consumed = (2.695 − 0.977) g
= 1.718 g
Amount of calcium in sample = (mass calcium)/($M$ Ca)
= 0.977 g/40.08 g mol$^{-1}$
= 0.0244 mole

Amount of chlorine in sample = (mass chlorine)/($M$ Cl)
= 1.718 g/35.45 g mol$^{-1}$
= 0.0485 mole

$$\frac{\text{Atoms Ca}}{\text{Atoms Cl}} = \frac{\text{moles Ca}}{\text{moles Cl}} = \frac{0.0244 \text{ mol}}{0.0485 \text{ mol}} = \frac{1.00}{1.99} = \frac{1}{2}$$

The empirical formula is thus $CaCl_2$.

## PROCEDURE

### SAFETY NOTES

Besides reacting with the metal to form a sulfide, sulfur also reacts when heated with the oxygen in the air to form gaseous sulfur dioxide, $SO_2$. This highly reactive gas destroys lung tissue when inhaled in only moderate concentration and is a dangerous irritant even in low concentration. For this reason, all heating of sulfur must be carried out in or under a fume hood, with a strong through draught. Before beginning any heating, check the hood air flow by closing the window of the hood and placing a tissue wipe across the top grille. An adequate through draught will hold the tissue in place on the grille. Keep the hood windows closed at all times, except when introducing or removing the crucible or when manipulating the burner. In this last case keep the window as low as possible. If you are using a canopy hood, your instructor will advise you on how to test it.

Handle hot crucibles only with tongs. Avoid placing hot crucibles on paper or other combustible materials. During the heating steps, the rings or tripod tops become very hot. Remember that hot metal takes some time to cool after the flame has been removed, so do not handle these items prematurely.

### 1. Labeling the Crucibles

Place a *small* distinguishing mark on each crucible and lid with the ferric chloride solution, using a wooden splint. Upon heating, the ferric chloride is converted to an iron oxide, which adheres firmly to the porcelain surface. Use of too much chloride will result in flaking of the oxide and lead to errors in weighing. Use the distinguishing marks to associate each crucible with the proper lid and to avoid confusion of samples.

### 2. Heating the Crucible Assembly to Constant Weight

On a tripod or ring stand support a clean dry crucible and lid on a clay triangle, as shown in Figure 3.1. Do not use wire gauze. Get a clear, hot blue flame on the burner and heat the crucible assembly to redness for about 5 minutes.

Allow the crucible assembly to cool to room temperature. With the crucible tongs remove the assembly to the fine balance and weigh it to the nearest milligram.

Reheat the assembly as before, allow it to cool and reweigh it once again to the nearest mg. Repeat this process until two successive weighings agree to within 10 mg.

### 3. Adding the Reagents

Using the rough balance, add the specified mass of metal to the crucible (about 2 g of copper or manganese, about 4 g of lead or tin). Use the fine balance to establish the mass of the crucible, metal, and lid to the nearest milligram. Remember not to add

# Determination of the Empirical Formula of a Metal Sulfide

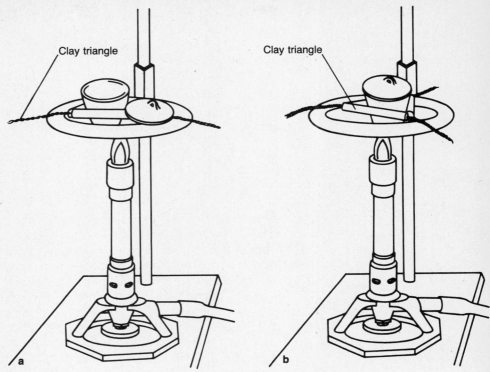

**Figure 3.1**
a Heating the crucible and lid to dryness using a ring stand and clay triangle.
b The crucible assembly ready for ignition. The lid is fully closed to minimize oxidation.

reagents directly to the crucible at the fine balances, which are sensitive precision instruments and easily damaged by introduction of foreign matter.

Once again at the rough balance, add about 1.5 g of sulfur to the crucible so as to cover the metal, and with a glass stirring rod mix the metal powder and sulfur together, taking care to avoid spillage from the crucible. In carrying out this step, first weigh out about 5 g of sulfur into a 50-ml beaker at the rough balance and use that sample for this and later required additions of sulfur. To minimize waste, transfer into the beaker all of the sulfur you tip from the reagent bottle into the bottle cap. Note that at no stage is it necessary to know the *exact* mass of sulfur added to the crucible, as an excess is always present.

## 4. Carrying Out the Reaction

Transfer the crucible assembly and contents to a stand with a clay triangle *set up in the fume hood*. Heat the crucible and contents *gently* with a small clear burner flame, playing the flame over the base and sides of the crucible. If blue flames of burning sulfur appear, stop the heating until they disappear. Now, with a strong flame, heat the crucible assembly for about 15 minutes. At high temperature the entire crucible should glow red hot. Recall that the hottest part of a bunsen burner flame is about 1 cm above the tip of the central blue cone. During the initial heating the elements combine to form the sulfide and this reaction continues at higher temperatures, although the stronger heating serves mainly to burn off excess sulfur as sulfur dioxide, $SO_2$. Be sure that the lid of the crucible remains firmly in place during the heating and that the lid is also heated in the latter stages of the process. If the lid remains cool and insufficient heat is applied, sulfur vapors may condense on the underside of the lid and make it difficult to remove. If the lid is displaced during heating, then oxides or sulfates may form as contaminants.

Allow the crucible assembly to cool to room temperature in the hood, then remove the lid and inspect the contents. If any sulfur remains unreacted (check under the lid) replace the lid and reheat the assembly red hot for another 5 to 10 minutes. If all the sulfur has reacted, then weigh the assembly and contents to the nearest milligram. (*Be sure to use tongs!*)

The reaction is rarely complete in a single heating, so further addition of sulfur and reheating is required.

Very carefully, taking care not to spill any of the contents, try to break up the sintered sulfide in the crucible. If it proves to be too hard to break up, ignore this step.

Next add about 1 g of sulfur to the crucible and contents and mix it well with the sulfide, again taking care to avoid spillage. Replace the crucible lid and repeat the heating, cooling, and weighing steps until two successive weighings agree to within 20 mg, at which point the reaction is assumed to be complete. If proper care is taken in the initial heating at moderate temperature at each stage, then only one or two extra additions of sulfur should be needed.

Repeat Steps 2 to 4 for a second sample of the same metal.

In carrying out the experiment you will work in pairs. Divide the work between you in whatever way seems most advantageous, but reverse your roles during the second run. Each student in a pair should record the data for the two independent determinations on his own data sheet and you may share this information freely between you, but each student should complete the report sheet independently. Do the calculations for both runs but report only the results for the better of the two.

Kurt Swanick

Name _____ Instructor _____

Partner _____

## DATA SHEET

Enter quantities neatly and directly with appropriate dimensions. Use a pen not a pencil and make no erasures. In case of error, cross out the incorrect entry and substitute the correct one above the deletion. Do not make preliminary entries on scraps of paper, etc. Use carbon paper to prepare the duplicate sheet.

|  | Assembly 1 | Assembly 2 |  |
|---|---|---|---|
| Mass of crucible + lid | ~~17.8~~ ~~17.9~~ | 18.778 g | (1) |
|  | 17.894 g | 18.777 g | (2) |
|  | _____ | _____ | (3) |
| Difference between final weighings | 0.003 g | 0.001 g |  |
| Mass of crucible + lid + metal | 21.756 g | 22.121 g / 24.582 g |  |
| Mass of crucible + lid + metal + sulfur | 24.031 g | ~~23.75~~ |  |
| Mass of crucible + lid + sulfide | 22.215 g | 22.596 g | (1) |
|  | 22.258 g | _____ | (2) |
|  | 22.213 g | 22.565 | (3) |
|  | _____ | _____ | (4) |
| Difference between final weighings | _____ | _____ |  |

85

Name _____ Instructor _____

Partner _____

## DATA SHEET

Enter quantities neatly and directly with appropriate dimensions. Use a pen not a pencil and make no erasures. In case of error, cross out the incorrect entry and substitute the correct one above the deletion. Do not make preliminary entries on scraps of paper, etc. Use carbon paper to prepare the duplicate sheet.

|  | Assembly 1 | Assembly 2 |  |
|---|---|---|---|
| Mass of crucible + lid | 17.896 g | 18.778 g | (1) |
|  | 17.894 g | 18.777 g | (2) |
|  | _____ | _____ | (3) |
| Difference between final weighings | 0.002 g | 0.001 g |  |
| Mass of crucible + lid + metal | 21.756 g | 22.121 g |  |
| Mass of crucible + lid + metal + sulfur | 24.081 g | 24.582 g |  |
| Mass of crucible + lid + sulfide | 22.215 g | 22.596 g | (1) |
|  | 22.213 g | 22.565 g | (2) |
|  | _____ | _____ | (3) |
|  | _____ | _____ | (4) |
| Difference between final weighings | 0.002 g | 0.031 g |  |

87

## WORKSHEET

Carry out all the calculations for Experiment 3 on this sheet, attach it to your report sheet, and submit the two together. Do your calculations directly on this sheet. Do not do them on scratch paper and then copy them on to this sheet. Make no erasures and delete incorrect entries with a single line so as to leave the original data legible. Use a pen.

① $\quad 21.756\,g$
$\quad\; -\; 17.894\,g$
$\quad\quad\; 3.862\,g \to m$ of Pb

② $\; 0.457\,g \to m$ of S

③ $\; \dfrac{3.862}{207.2} = 0.01864$

④ $\; \dfrac{2.325}{32.06} = 13.79$

⑤ $\; 0.01864 : 13.79$

⑥ ~~1 : 740~~

⑦ $\; 1 : 740 \; ; \; Pb\,S_{740}$

# EXPERIMENT 4

# Synthesis and Purification of Potassium Bromide and Bromate

### AIM

To synthesize potassium bromide, KBr, and potassium bromate, $KBrO_3$, by direct reaction of liquid bromine with an aqueous solution of potassium hydroxide, and to separate and purify these salts by fractional crystallization and recrystallization.

### EQUIPMENT

| | |
|---|---|
| From stockroom: | Buchner filter funnel and flask |
| | Crucible and lid. |
| From locker: | 200, 250 or 300 mL Erlenmeyer flask |
| | Evaporating dish |
| | Thermometer |
| | Graduated cylinder |
| | Wash bottle. |
| Laboratory: | Bunsen burner |
| | Tripod stand and gauze |
| | Ice bath |
| | Filter paper. |

## REAGENTS

Potassium hydroxide pellets, KOH
Liquid bromine, $Br_2$ (to be dispensed by instructor)
$3M$ sulfuric acid, $H_2SO_4$
$3M$ nitric acid, $HNO_3$
$0.1M$ silver nitrate, $AgNO_3$
Acetone and dichloromethane
Ice for cooling

## BACKGROUND

### Associated Readings

The following sections of *Chemical Principles* should be read in conjunction with Experiment 4.

**Chapter 1** The Science of Chemistry
        1.2 Chemistry as a Physical Science
**Chapter 2** The Language of Chemistry
        2.3 Chemical Names and Formulas
        2.4 Balanced Chemical Equations
        2.5 Important types of Chemical Reactions
        2.6 States of Substances
**Chapter 3** Stoichiometry
        3.2 Mass Relationships in Chemical Formulas
            (The Limiting Reagent)
        3.3 Mass Relationships in Chemical Reactions

Optional reading for additional background:

**Chapter 7** The Chemical Bond
        7.6 Oxidation Numbers
**Chapter 9** The Chemistry of the Nonmetals
        9.7 The Halogens (Halogen–Oxygen Compounds)

This experiment illustrates some aspects of the aqueous solution chemistry of bromine, a chemistry it shares with the other halogens, chlorine and iodine.

Bromine, $Br_2$, like the other halogens, is moderately soluble in water. A saturated solution of liquid bromine in water has about a $0.2M$ concentration of the dissolved element. But bromine in aqueous solution reacts with water to form bromide ion and hypobromous acid:

$$Br_2(aq) + H_2O \rightleftharpoons H^+(aq) + Br^-(aq) + HOBr(aq) \tag{1}$$

Such a reaction, in which a substance in a given oxidation state is converted to two separate species, one of which is in a lower and the other in a higher oxidation state, is known as a disproportionation reaction. In this case elemental bromine is in oxidation state 0, bromide ion has bromine in oxidation state $-1$, and hypobromous acid has bromine in oxidation state $+1$. You may not yet be familiar with the term "oxidation state." If so, simply ignore any references to the term in this background material. If, however, you are familiar with the term, you may wish to review Section 7.8 of *Chemical Principles*, where it is discussed in detail.

As equation (1) implies, hypobromous acid (HOBr) is a weak acid that is not dissociated to any significant extent in neutral or acidic solution, whereas hydrobromic acid (HBr) is a strong acid, and thus exists in aqueous solution in an effectively completely dissociated state.

At room temperature, reaction (1) proceeds only to a very limited extent, so that the concentrations of bromide ion and hypobromous acid in solution are very small. An aqueous solution of bromine thus consists almost exclusively of dissolved bromine as long as the solution remains either acidic or neutral.

In basic solutions containing appreciable concentrations of the hydroxide ion, $OH^-$, the concentrations of the species in (1) are profoundly altered. The hydrated proton and the hypobromous acid both react rapidly and completely with hydroxide ions in neutralization reactions:

$$H^+(aq) + OH^-(aq) \longrightarrow H_2O \qquad (2)$$

$$HOBr(aq) + OH^-(aq) \longrightarrow H_2O + BrO^-(aq), \qquad (3)$$

so that the overall reaction of bromine in basic or alkaline solution is

$$Br_2(aq) + 2OH^-(aq) \longrightarrow Br^-(aq) + BrO^-(aq) + H_2O \qquad (4)$$

However, at other than very low temperatures, the hypobromite ion, $BrO^-$, is itself unstable, disproportionating to give bromide ion and bromate ion, $BrO_3^-$ (oxidation state $+5$), according to the equation

$$3BrO^-(aq) \longrightarrow 2Br^-(aq) + BrO_3^-(aq) \qquad (5)$$

This reaction goes effectively to completion and at temperatures above 50°C is very fast. A hot alkaline solution of bromine thus contains $Br^-$ and $BrO_3^-$ ions as the only bromine-containing species, and if the base chosen is potassium hydroxide, as in this experiment, the solution will yield both potassium bromide and potassium bromate.

These two salts differ markedly in their solubilities in water at different temperatures, as illustrated in Figure 4.1, and we use these differences to separate the two. Figure 4.1 shows that at temperatures around 0°C almost all of the potassium bromate may be crystallized from solution with little contamination from potassium bromide. We use this fractional crystallization procedure to separate the two salts, crystallizing out the potassium bromate at near 3°C and evaporating the residual solution to crystallize potassium bromide.

Each salt will be contaminated with small amounts of the other. We purify the bromate by recrystallization from water and purify the bromide by heating the sample,

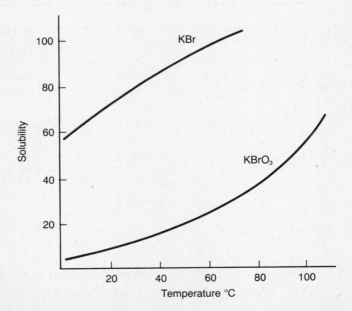

**Figure 4.1**
Solubility of KBr and $KBrO_3$ (g/100 g of water) at various temperatures.

noting that at high temperatures potassium bromate decomposes to yield potassium bromide and oxygen. (Compare the decomposition of $KClO_3$ to yield $KCl$ and $O_2$ in Experiment 5.)

$$2KBrO_3(s) \longrightarrow 2KBr(s) + 3O_2(g) \tag{6}$$

## PROCEDURE

### SAFETY NOTES

A high degree of discipline must be observed in the laboratory during this experiment. Bromine, in liquid or gaseous form, is a highly corrosive toxic agent. Liquid bromine in contact with skin produces painful blistering burns, which heal only slowly. Bromine vapors rapidly and thoroughly dissolve lung tissues if inhaled, and excessive inhalation of the vapor is fatal. It is thus of the utmost importance that you treat this powerful reagent with the greatest care and respect and obey to the letter all instructions for its safe use. You will only handle bromine in basic solution, where it is relatively harmless.

However, you should not carry the flask containing the bromine solution into the main area of the laboratory, but take it immediately to the hood assigned to you. Do not remove the flask from the hood until instructed to do so by the instructor.

Should you accidentally spill bromine on your skin, immediately flush it off with copious amounts of cold running water, continuing to wash the affected area with cold water for at least 15 minutes. Any spill, other than a very slight one, is likely to require skilled treatment. Immediately notify your instructor of any accident involving bromine, no matter how trivial it may seem.

**If the procedures outlined below are followed, there is no danger to you or others from the use of bromine in this experiment.**

Potassium hydroxide pellets are hygroscopic, picking up water from the air to form corrosive puddles of base. Do not handle the pellets. Clean up any spills immediately, depositing waste pellets in the sink and flushing with lots of water. Do not leave the reagent bottle containing the KOH pellets open, but replace the bottle cap immediately after use. Do this even if someone is waiting directly in line behind you; it is all too easy for the twelfth person in line to forget to replace the cap because it was not there when he started using the bottle.

**Acetone is volatile and flammable. Avoid exposing the vapors to flames.**

### 1. Preparation of Potassium Hydroxide Solution

Using the rough balance, weigh out about 10 g of KOH pellets into a clean, dry 200-mL Erlenmeyer flask. Add about 60 mL of distilled water to the pellets in the flask, swirling the flask gently to help dissolve the solid. When solution is complete, cool the flask and its contents to room temperature by holding under the cold water tap for a few minutes. Check the temperature at the end of this cooling period, being sure that it is below 25°C before proceeding further. Rinse off the thermometer with distilled water after withdrawing it from the flask.

### 2. Preparation of an Ice Bath

Fill a polystyrene container one-third full with crushed ice. Add about 250 mL of cold tap water to the container to make an ice-water mix. This mix is a more efficient cooling medium than ice alone because it allows more of the surface of the flask to be in contact with the cooling medium. Fill your wash bottle with distilled water and place it in the ice bath to cool.

### 3. Addition of Bromine

Strict discipline is required during this step. Read through this section twice to make sure you completely understand the procedure to be followed.

Dry off the outside of the flask containing the cooled hydroxide solution and take it to the central hood where your instructor will add 5 mL of liquid bromine to the flask and stopper it with a cork. Under the direction of your instructor take the stoppered flask to one of the previously prepared burner-stand assemblies in another hood. Holding the stoppered flask inside the hood, very gently swirl the contents of the flask to promote mixing and reaction. When the solution is a uniform orange color, place the flask on a wire gauze on one of the stands. Wait till all students using the hood have completed this step, remove the stopper from the flask and bring the contents to a gentle boil with a burner. Continue to boil the solution until the color changes to a very pale straw shade (the color of dry white wine is a good approximation). Ask your instructor to verify that you have indeed removed all excess bromine, extinguish the burner, and carefully carry the flask and its contents to your station. Do this by carrying the flask with your fingers at the rim, supporting the base, if necessary, on a towel held in your other hand.

### 4. Isolation of Potassium Bromate

Cool the flask to room temperature by again holding it under the cold water tap. Check the solution volume in the flask; the boiling process should not have resulted in the loss of more than about 15 mL of your original 60 mL. If excessive evaporation has taken place, make up the volume by addition of an appropriate quantity of distilled water, but take care not to add too much.

Remove the wash bottle from the ice bath and place the flask in the bath. It may be necessary to add a little more ice at this point. Allow the contents of the flask to cool to about 3°C, at which temperature crystals of the bromate should separate out. Crystallization may be promoted by scratching the inside walls of the flask with a glass stirring rod.

While the solution is cooling, assemble the Buchner filter funnel and flask, securing them with a stand and clamp. Check the aspirator action by moistening a filter paper disc with water and placing it on the filter bed. With the aspirator on, the filter paper should be sucked down onto the bed.

When crystallization of the bromate is complete, decant off the bulk of the liquid above the crystals in the flask (the supernatant liquid) through the filter. Transfer the crystalline slurry onto the filter with the help of a glass rod and rubber policeman and with the glass rod spread the crystals out over the filter bed. With the bottom of a small beaker, press the crystals gently onto the filter paper to help remove attached residual solution. Wash the crystals twice with *small* amounts of ice-cold distilled water. Continue the aspiration for a few more minutes then remove the filter paper and crystals from the funnel. Press the crystals between two fresh filter papers to further dry them, transfer them to a previously weighed 50-mL beaker and weigh to the nearest 10 mg on a rough balance to establish the yield of crude product.

### 5. Isolation of Potassium Bromide

The potassium bromide is isolated by evaporating the filtrate from the previous step in an evaporating dish. The total volume of the filtrate will likely exceed the volume of your evaporating dish, so adopt the following procedure.

Balance the evaporating dish carefully on a clay triangle atop a tripod stand. Add filtrate to the dish until it is no more than one-third full, pouring from the flask with a glass stirring rod, used as shown in Figure 1.1, to avoid dripping. Apply a gentle blue-flame heat to the base of the dish to reduce the volume of solution to about 10 mL. Top up the evaporating dish with filtrate to the one-third level and repeat the process until all of the filtrate has been transferred to the dish. Toward the end of the process a sludge of bromide will begin to form and spread up the sides of the dish. Push it down into the body of the solution as it forms, using a glass rod. Do not rush this step, do not overfill the evaporating dish, and do not boil the solution. Evaporation takes place quite effectively below the boiling point and boiling or overfilling will only lead to sputtering and loss of product.

As the evaporation is proceeding, clean out the filter funnel and flask assembly and make it ready once more. When the sludge in the evaporating dish has been reduced to a total volume of about 5 mL, but is still gooey, transfer it into the funnel, pressing it onto the filter paper as was done for the bromate. When the water has been drawn off into the flask, wash the bromide with about 10 mL of acetone and continue aspiration until the odor of acetone is no longer noticeable above the filter bed. Remove the crude product from the filter and weigh it on a rough balance to the nearest 10 mg. Record the mass of product.

## 6. Purification of Potassium Bromate

Refer to the solubility graph in Figure 4.1 and determine the minimum volume of water needed to dissolve the bromate preparation at 80°C. Add this volume of water to the crystals in a 100-mL beaker and heat until all of the solid has dissolved. Allow the solution to cool to room temperature and then place the beaker in the ice bath to cool to 3°C until crystallization is complete.

Collect the recrystallized materials as before, washing the crystals on the filter bed with 10 mL of *ice-cold* distilled water and drying with 10 mL of acetone. Continue the aspiration until the odor of acetone is no longer apparent, then remove and weigh the purified product to the nearest 10 mg. Record the yield.

The recrystallized bromate should now be tested for purity. Take a small quantity of the bromate (a chemist's smidgeon) in a test tube and dissolve it in 2 mL of distilled water to which have been added 3 drops of $3M$ nitric acid, $HNO_3$. Heat this solution and add a drop of $0.1M$ silver nitrate. If any bromide is present, a pale yellow precipitate of silver bromide will form. If the sample is pure, the solution will remain clear. If you have carried out the procedure well, a single recrystallization will yield a pure product. Show your instructor the result of this test and have him initial your data sheet entry describing the result. If the product is impure and if time permits, run a second recrystallization.

## 7. Purification of Potassium Bromide

In acid solution, bromate anion combines with bromide to give bromine:

$$BrO_3^-(aq) + 5Br^-(aq) + 6H^+(aq) \longrightarrow 3Br_2(aq) + 3H_2O \qquad (7)$$

This reaction may be used as a test of the purity of your sample. Add a small quantity of the preparation to 3 mL of $3M$ sulfuric acid. Add a few drops of dichloromethane to the mix and shake the test tube. The development of a yellow color in the organic layer indicates bromine formation and hence contamination by bromate.

To remove this contaminant we use the thermal decomposition reaction by which potassium bromate is converted to potassium bromide with liberation of oxygen.

$$3KBrO_3(s) \longrightarrow 3KBr(s) + 3O_2(g) \qquad (8)$$

Weigh the crucible to the nearest 10 mg on a rough balance and add to it between 1.5 and 2.0 g of the crude bromide, noting to the nearest 10 mg the amount added.

Support the crucible on a clay triangle atop a tripod stand and heat the crucible and its contents for about 15 minutes. The crucible should be heated gently at first and finally to red heat. Allow the crucible and its contents to cool to room temperature and reweigh to establish the new mass of pure product. Test the purified material to check that all detectable bromate has been removed.

Name _____  Instructor _____

## PRELABORATORY PREPARATION SHEET

1. Name the following bromine containing species:

   $BrO^-$ _____     $BrO_3^-$ _____

   $Br^-$ _____     $HOBr$ _____

2. A disproportionation reaction is one in which a species in a particular oxidation state is converted to two products, one of which is in a higher, the other in a lower, oxidation state than the original. Two such reactions are given in the background material. Identify them and record them here.

   _____

   _____

3. Find in the background material an example of a reverse disproportionation reaction and record the equation here.

   _____

4. Derive from Equations (2) and (3) in the background material, the overall reaction between bromine and potassium hydroxide in aqueous solution. Record the equation here.

   _____

5. Given the quantities of each to be used in the experiment, which is the limiting reagent in this reaction? Show your calculation. (Density of $Br_2(l)$ = 3.12 g/cm$^3$.)

   _____

   _____

   _____

   _____

6. About what mass of KBr is expected to result from this experiment? Show your calculation.

   _____

   _____

   _____

   _____

7. Determine the approximate mass of $KBrO_3$ expected. Show your calculation.

   _____

   _____

   _____

   _____

8. In the filtration of the recrystallized KBrO₃, and in the filtration of KBr, acetone is used to dry the crystals in the filter bed. Why may this procedure not be adopted in the filtration of the crude KBrO₃ preparation?

Name _____  Instructor _____

## DATA SHEET

Mass of KOH used: _____

Volume of $Br_2$ added: _____

1. Mass of $KBrO_3$ from initial crystallization: _____
   Test for purity (check one):

   Negative _____  Positive _____  Instructor's OK _____

2. Mass of $KBrO_3$ from first recrystallization: _____
   Test for purity (check one):

   Negative _____  Positive _____  Instructor's OK _____

3. Mass of $KBrO_3$ from second recrystallization: _____
   Test for purity (check one):

   Negative _____  Positive _____  Instructor's OK _____

4. Mass of crude KBr: _____
   Test for purity (check one):

   Negative _____  Positive _____  Instructor's OK _____

5. Mass of KBr taken for purification: _____

6. Mass of pure KBr: _____
   Test for purity (check one):

   Negative _____  Positive _____  Instructor's OK _____

Name _____  Instructor _____

## DATA SHEET

Mass of KOH used: _____

Volume of $Br_2$ added: _____

1. Mass of $KBrO_3$ from initial crystallization: _____
   Test for purity (check one):

   Negative _____   Positive _____   Instructor's OK _____

2. Mass of $KBrO_3$ from first recrystallization: _____
   Test for purity (check one):

   Negative _____   Positive _____   Instructor's OK _____

3. Mass of $KBrO_3$ from second recrystallization: _____
   Test for purity (check one):

   Negative _____   Positive _____   Instructor's OK _____

4. Mass of crude KBr: _____
   Test for purity (check one):

   Negative _____   Positive _____   Instructor's OK _____

5. Mass of KBr taken for purification: _____

6. Mass of pure KBr: _____
   Test for purity (check one):

   Negative _____   Positive _____   Instructor's OK _____

Name _____ Instructor _____

## REPORT SHEET

1. From the actual mass of KOH pellets used, calculate the expected yield of KBr and $KBrO_3$. Show your calculations with proper units.

   _____

   _____

   _____

   _____

   Expected mass of KBr: _____ Expected mass of $KBrO_3$: _____

2. Give your yield of crude $KBrO_3$ as a percentage of the theoretical yield. Show your calculation.

   _____

3. Give your yield of purified $KBrO_3$ as a percentage of the theoretical yield. Show your calculation.

   _____

4. Give your yield of crude KBr as a percentage of the theoretical yield. Show calculation.

   _____

5. Mass of KBr taken for purification: _____

6. Mass of purified KBr: _____

7. Percent yield in purification process: _____

8. Overall percent yield of KBr for purified sample: _____

## WORKSHEET

Carry out all the calculations for Experiment 4 on this sheet, attach it to your report sheet, and submit the two together. Do your calculations directly on this sheet. Do not do them on scratch paper and then copy them on to this sheet. Make no erasures and delete incorrect entries with a single line so as to leave the original data legible. Use a pen.

# EXPERIMENT 5

# Determination of the Standard Molar Volume and the Gas Constant, R

### AIM

To find the molar volume of oxygen and the value of the gas constant R, by measurement of the volume of gas corresponding to a known mass of oxygen at determinate conditions of temperature and pressure.

### EQUIPMENT

| | |
|---|---|
| From stockroom: | 500-mL round-bottomed flask |
| | Rubber tubing |
| From locker: | Test tube |
| | 1-L beaker |
| | Rubber stoppers |
| | Pinchcock clamp |
| Laboratory: | Stand and clamp |
| | Glass tubing and file |
| | Burner |
| | 500-mL graduated cylinder |

## 114 EXPERIMENT 5

## REAGENTS AND OTHER MATERIALS

Potassium chlorate, $KClO_3$
Manganese dioxide catalyst, $MnO_2$
Glycerin lubricant

## BACKGROUND

### Associated Readings

The following sections in *Chemical Principles* should be read in conjunction with Experiment 5.

**Chapter 3** Stoichiometry
               3.3 Mass Relationships in Chemical Reactions
**Chapter 4** States of Matter
               4.2 Vapor Pressure
               4.3 The Gaseous State
               4.4 The Ideal Gas Equation
               4.5 Dalton's Law of Partial Pressure

See also Example 4.14.

Two laws relating to the physical properties of gases are

1. **Boyle's law**—the volume of a fixed mass of gas at a fixed temperature is inversely proportional to the gas pressure.
2. **Charles's law**—the volume of a fixed mass of gas at a fixed pressure is directly proportional to the absolute temperature of the gas.

These two laws may be expressed mathematically as

$$V = k_1/P$$

or

$$PV = k_1$$

and

$$V = k_2 T$$

where $V$ is the volume, $P$ is the pressure, and $T$ is the temperature.

Combination of these relationships (see Section 4.3 of *Chemical Principles*) yields

$$\frac{PV}{T} = k,$$

so that for a given gas sample, where any two measurements of $P$, $V$, and $T$ must yield the same value of the constant, $k$, we may write

$$\frac{P_1 V_1}{T_1} = \frac{P_2 V_2}{T_2}$$

where the subscripts distinguish the two sets of measurements. Obviously, if any five of the six terms are known, the value of the unknown sixth is readily obtainable as, for example:

$$V_2 = V_1 \times (T_2/T_1) \times (P_1/P_2)$$

A most important interpretation of Charles's law was made by Amadeo Avogadro. Noting that *all* gases expand to the same extent with an equal rise in temperature, Avogadro reasoned that under the same conditions of temperature and pressure, equal volumes of different gases must contain the same numbers of particles and that these particles need not be individual atoms but might be combinations of atoms—what we now call molecules. Acceptance of the idea that many elemental gases existed as diatomic molecules, coupled with Avogadro's hypothesis, allowed Stanislao Cannizzaro to reconcile conflicts between differing contemporary atomic weight scales. Cannizzaro's contribution was instrumental in starting D. I. Mendeleev on the way to the Periodic Table.

An important quantity in these discussions is the volume occupied by one mole of a gas under standard conditions of temperature and pressure, the so-called **standard molar volume.** Standard conditions, **(STP),** are defined as exactly 1 atmosphere pressure and exactly 0°C (273 K). This volume, the volume occupied by Avogadro's number of molecules, is obviously the same for all gases if the hypothesis holds.

One purpose of Experiment 5 is to establish the standard molar volume of oxygen by measuring the volume of gas generated under known conditions of temperature and pressure from a known mass of potassium chlorate. At temperatures above 400°C this substance decomposes, yielding oxygen gas and potassium chloride according to the equation

$$2KClO_3(s) \longrightarrow 2KCl(s) + 3O_2(g)$$

The mass of oxygen generated is found by subtraction of the mass of solid product from that of solid reactant, the difference, by the **law of conservation of mass,** being the mass of gas produced in the reaction. The volume of gas produced may be measured by the displacement of an equivalent volume of water at the barometric pressure prevailing in the laboratory, and the temperature of the gas may be directly measured with a thermometer.

The equation $PV/T = k$ is conveniently rearranged as

$$PV = kT$$

or as

$$PV = nRT$$

where $n$ is the amount of gas in moles and $R$ is now a constant common to all gases, the so-called **universal gas constant.** With $P$, $V$, $n$, and $T$ known for a particular gas sample, $R$ may be evaluated. In this experiment we will accept the molecular weight of oxygen as 32.00, so that $n$ is known, and we will evaluate $R$ and compare our results with the accepted value of 0.082057 L atm/mol K.

The measurement of gas pressure in this experiment, where the oxygen gas is collected over water, requires consideration of a third law of gas behavior:

**3. Dalton's law of partial pressures**—the total pressure exerted by a mixture of gases is the sum of the partial pressures exerted by each component of the mixture.

**Partial pressure** is defined as the pressure which would be exerted by a particular component if it were the sole gaseous species occupying the volume in question. In this experiment, because the oxygen is confined over water, we actually have two gaseous components present in the flask used to collect the oxygen generated. Besides oxygen gas, the flask is also saturated with water vapor whose partial pressure is the vapor pressure of water at the temperature of the water in the flask. The total pressure, which may be made equal to the barometric pressure of the laboratory, is then

$$P_{total} = P_{O_2} + P_{H_2O}$$

and the quantity of interest in our evaluation is $P_{O_2}$, the pressure attributable to oxygen alone.

**TABLE 5.1** Table of Water Vapor Pressures at Various Temperatures

| Temperature (°C) | Pressure (mmHg) | Temperature (°C) | Pressure (mmHg) |
|---|---|---|---|
| 20 | 17.54 | 26 | 25.21 |
| 21 | 18.65 | 27 | 26.74 |
| 22 | 19.83 | 28 | 28.35 |
| 23 | 21.07 | 29 | 30.04 |
| 24 | 22.38 | 30 | 31.82 |
| 25 | 23.76 | | |

The vapor pressure of water in the range of temperatures between 20°C and 30°C is given in Table 5.1. See also Table 4.1 in Section 4.5 of *Chemical Principles*.

## PROCEDURE

### SAFETY NOTES

The evolution of oxygen from $KClO_3$ may lead to an explosive reaction if certain reducing agents are present. Be sure that the test tube used for the reaction is scrupulously clean and dry before adding $KClO_3$ to it. Be sure that none of the glycerin used as a lubricant for your glass tubing is allowed to enter the test tube or a dangerous mixture may be formed.

Oxygen is generated under pressure in the experiment. Have your instructor check your apparatus before you turn on the burner so as to be sure there are no constrictions or clamps which might block gas flow.

Oxygen is very reactive. Keep all flames or naked lights away from the flask containing the gas.

Take care in handling the glass tubing, particularly when inserting the tubing into stoppers. Lubricate the tubing before inserting it and never use undue force. Keep your hand as close to the stopper as possible when inserting the glass tubing.

### 1. Preparing the Flask and Beaker

Cut three pieces of glass tubing, each 8 cm in length, from a longer piece of tubing. Fire-polish the ends of these pieces and allow them to cool. Avoid overheating, which may cause the ends to collapse and lead to a constriction in the gas flow. Cut and prepare a fourth length of tubing to form the L-bend shown in the illustration of the apparatus in Figure 5.1. There should be no constriction of the tubing at the bend.

Fill the round-bottomed flask with water to within a few centimeters of the top and add about 5 cm of water to the beaker. Place the glass and rubber tubing inserts into the flask and beaker as shown in Figure 5.1, but do not connect the test tube assembly at this stage. Be sure that the L-bend comes close to the bottom of the flask; it is important that the bottom of the tube remain under water throughout the experiment. Equally, be sure that the entry tube to the flask from the reaction tube remains above water at all times. Fill the tubing connecting the flask and beaker with water by gently blowing into the entry tube until water flows steadily from the flask to the beaker. Siphon water to and fro from the beaker to the flask until all air bubbles have been cleared from the tubing. Now refill the flask with water until the level is just below the end of the entry tube. Close the clamp on the entry tube firmly.

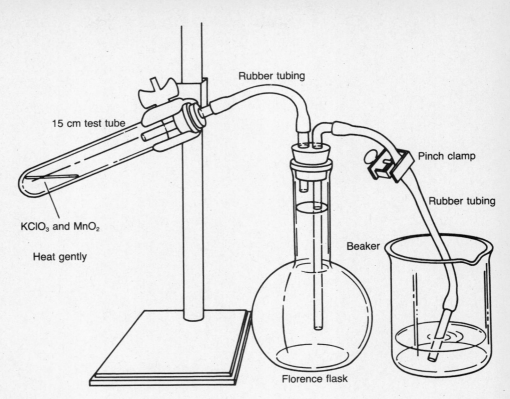

**Figure 5.1**
Apparatus for oxygen generation and collection of displaced water.

## 2. Setting up the Test Tube Assembly

Place the test tube to be used for the reaction in a clean, dry 250- or 100-mL beaker and on the rough balance add about 3 g of potassium chlorate to the test tube.

Heat the bottom of the test tube gently over a clear blue flame until the potassium chlorate just melts, but not so much as to produce any effervescence. When melting is complete, remove the test tube from the flame and allow the tube and its contents to return to room temperature. This step ensures that any water present in the solid is driven off.

When the test tube has cooled, replace it in the same clean, dry beaker and, by visual estimation, add about 10 mg of manganese dioxide to the test tube to act as a catalyst in the reaction. Weigh the whole assembly to the nearest mg on the fine balance, taking care that the tube does not snag the sides of the draught tower on the balance.

Support the test tube in a stand and clamp as shown in Figure 5.1, and connect it to the entry tube to the flask. Make sure no glycerin gets into the test tube during this step. Check that the entire apparatus is airtight by opening the clamp on the exit tube. No more than a drop or two of water should flow from the flask.

Equalize the external and internal pressures by raising the beaker manually until the water level in it is the same as that in the flask. Close the exit tube clamp, remove the beaker, discard any water it contains, dry it out, and replace the exit tube so that its end rests on the bottom of the beaker. Open the clamp on the exit tube. Once again, if the pressures have been equalized and the vessel is airtight, no water should flow.

Before proceeding, have your instructor check your apparatus to make sure that it is correctly and safely set up. Do not go forward until this check has been made.

## 3. Generation of Oxygen

With a small clear-blue flame on the burner, gently warm the $KClO_3$ and $MnO_2$ mixture in the test tube. Distribute the heat evenly over the lower part of the test tube and do not increase the heat until the mixture has melted and a gentle evolution of gas has begun to displace water from the flask to the beaker. Pure oxygen is invisible; if cloudy white fumes form, stop the heating at once and inform your instructor. Once a steady evolution of oxygen has begun, you may increase the heat applied to the test tube. Continue heating until there is no further evolution of oxygen or until about 300 mL of water have been displaced, whichever comes first. In any event do not allow the water level in the flask to fall below about 3 cm from the end of the exit tube.

## 4. Final Measurements

Remove the flame from the test tube, extinguish the burner, and allow the assembly to return to room temperature. When you judge the cooling to be complete, equalize the internal and external pressures as before and close the exit tube clamp. Remove the beaker and its contents and record the temperatures of both the gas and water in the flask. Measure the volume of displaced water with a 500-mL graduated cylinder. Disconnect the test tube and weigh it and its contents in the same beaker as before, once again to the nearest milligram.

Record the raw barometric pressure in the laboratory and apply any necessary correlations, following the instructions posted next to the barometer. From the table of water vapor pressures find the partial pressure of water applicable to your experiment.

Repeat the experiment if time permits. If you have been working as a team, reverse the roles of the team members.

Name _____  Instructor _____

## PRELABORATORY PREPARATION SHEET

1. Give the equation for the decomposition of $KClO_3$ to yield oxygen.
   _____

2. What role is played by the manganese dioxide in the above reaction?
   _____

3. Why is it necessary to wipe excess glycerin from the test tube stopper and glass tubing before they are inserted into the test tube?
   _____

4. What is meant by partial pressure?
   _____
   _____

5. Why must the internal pressure in the apparatus and the external pressure in the laboratory be equalized?
   _____
   _____

6. What would be the effect of leaving the clamp on the exit tube closed while heating the $KClO_3$ mixture?
   _____

7. What kinds of correction must be made to barometric pressures read from the laboratory barometer?
   _____
   _____

8. Assuming normal conditions of temperature and pressure in the laboratory, roughly what mass of oxygen gas is to be expected from 3 g of potassium chlorate, and to what volume does this correspond? Show your calculation.
   _____
   _____
   _____
   _____

Name _____  Instructor _____

Partner _____

## DATA SHEET

1. Test tube assembly:

   Mass before reaction　　_____　　_____

   Mass after reaction　　_____　　_____

   Mass of $O_2$ generated　　_____　　_____

2. Temperature measurements:

   Gas in flask　　_____ °C　　_____ °C

   　　　　　　　　_____ K　　_____ K

   Water in flask　　_____ °C　　_____ °C

3. Pressure measurements:

   Uncorrected barometric pressure　　_____　　_____

   Corrections　　_____　　_____

   Corrected pressure　　_____　　_____

   Vapor pressure $H_2O$　　_____　　_____

4. Volume measurements:

   Volume of displaced water　　_____　　_____

   　　　　　　　　　　　　　　_____　　_____

Name _____  Instructor _____

Partner _____

## DATA SHEET

1. Test tube assembly:

    Mass before reaction    _____  _____

    Mass after reaction     _____  _____

    Mass of $O_2$ generated  _____  _____

2. Temperature measurements:

    Gas in flask      _____ °C   _____ °C

                      _____ K    _____ K

    Water in flask    _____ °C   _____ °C

3. Pressure measurements:

    Uncorrected barometric pressure  _____  _____

    Corrections                      _____  _____

    Corrected pressure               _____  _____

    Vapor pressure $H_2O$            _____  _____

4. Volume measurements:

    Volume of displaced water    _____  _____

                                 _____  _____

Name _____  Instructor _____

Partner _____

# REPORT SHEET

All calculations involved in this report are to be done on the worksheet for this experiment and should be attached to the Report Sheet on submission.

1. Enter from your Data Sheet:

   |  | Run 1 | Run 2 |
   |---|---|---|
   | Volume of $O_2$ gas | _____ | _____ |
   | Temperature of $O_2$ gas | _____ | _____ |
   | Pressure of $O_2$ gas | _____ | _____ |

2. Using the relationship $P_1V_1/T_1 = P_2V_2/T_2$, calculate the corresponding volume of gas at STP:

   _____  _____

3. Enter the mass of oxygen generated in each case:

   _____  _____

4. Calculate the number of moles oxygen from Step 3:

   _____  _____

5. From the entries in Steps 2 and 4 find the standard molar volume of oxygen:

   _____  _____

6. From the answer to Step 5, and assuming standard conditions, use the relation $PV = nRT$ to calculate the value of $R$. Give units:

   _____  _____

7. Check your answer to Step 6 by solving the equation $PV = nRT$ for $R$ by using your entries for Steps 1 and 4.

   _____  _____

8. The accepted values for standard molar volume and the universal gas constant are:

   Standard Molar Volume = 24.414 liter

   $R = 0.082057$ liter atm/mol K

   Express the difference between your calculated values and the accepted values as a percentage:

   Standard Molar Volume _____ %   _____ %

   $R$ _____ %   _____ %

125

9. Assuming that oxygen behaves as an ideal gas under the conditions of the experiment you have conducted (it does so quite closely), discuss the deviations found in Step 8 in terms of the uncertainties in the various experimental measurements. Reason quantitatively.

_____
_____
_____
_____
_____
_____

## WORKSHEET

Carry out all the calculations for this Experiment 5 on this sheet, attach it to your report sheet, and submit the two together. Do your calculations directly on this sheet. Do not do them on scratch paper and then copy them on this sheet. Make no erasures and delete incorrect entries with a single line so as to leave the original data legible. Use a pen.

# EXPERIMENT 6

# Determination of Molar Mass for a Volatile Liquid by Dumas's Method

**AIM**

To determine the molar mass of a volatile liquid by measurement of its vapor density

**EQUIPMENT**

From locker: Two Erlenmeyer flasks
1-L beaker
Laboratory: Ring stand and clamp
Burner and wire gauze

**REAGENTS AND MATERIALS**

Unknown volatile liquids and dispensing syringes
Aluminum foil and rubber bands
Distilled water

## BACKGROUND

### Associated Readings

The following sections of *Chemical Principles* should be read in conjunction with Experiment 6.

**Chapter 4** States of Matter
               4.2 Vapor Pressure
               4.3 The Gaseous State
               4.4 The Ideal Gas Equation
               4.5 Dalton's Law of Partial Pressure

Avogadro's hypothesis that equal volumes of gases under the same conditions of temperature and pressure contain equal numbers of molecules implies that the molecular weight of an unknown gas or vapor can be found by measurement of its vapor density. Stated in another way, the mass of a known volume of gas, measured under known conditions of temperature and pressure, is sufficient for a calculation of the molar mass.

The ideal gas equation of state is

$$PV = nRT$$

$P$ being the pressure, $V$ the volume, and $T$ the absolute temperature of the gas. $R$ is the gas constant and $n$ the amount of gas in moles. The amount of gas, $n$, may be expressed as

$$n = \frac{\text{mass of gas}}{\text{molar mass of gas}}$$
$$= m/\mathcal{M}$$

The unknown molar mass, $\mathcal{M}$, is then given by the ideal gas equation in the rearranged form

$$\mathcal{M} = mRT/PV$$

If the vapor density of the gas—i.e., the mass of gas per unit volume—is known under established conditions of temperature and pressure, $\mathcal{M}$ is readily found.

A simple but effective way of carrying out a molar mass determination for a volatile liquid is attributed to the French chemist J. Dumas. A quantity of the liquid is placed in a previously weighed glass container having a pinhole aperture. The liquid is heated and completely volatilized at a known temperature. Its vapors fill the container completely, driving out through the pinhole the air initially present. The container is then cooled to room temperature so that the vapors condense and air reenters the flask through the pinhole. The container is reweighed to find the mass of liquid and vapor remaining, i.e., the mass of vapor originally filling the flask. A correction is made for the mass of included air. The volume of the flask is derived from the mass of the water needed to fill it completely. The modified equation of state is then used to find the unknown molar mass.

Although accurate results are only possible by this method if a fairly carefully designed container is used, reasonable results can be obtained with a very simple apparatus and proper care in the experimental work. Typically, the method might be used to find a rough molar mass and a more accurate value found by combining the result with a carefully derived empirical formula for the compound. (See Section 3.2 of *Chemical Principles*.)

The determination of the mass of the condensed liquid and residual vapor is not as simple as it might seem. When the container is weighed "empty," the measured mass is actually that of the container plus that of the air it contains. Normally we disregard the mass of the air in a container because it is so small compared to the mass of any solid or liquid that displaces it. In this experiment, however, part of the air originally present is displaced by vapor of the unknown, a substance of comparable mass. Consequently, the difference in the masses of vapor and displaced air must be taken into account.

The mass of included air, $m_\alpha$, in the flask in the initial weighing may be found by applying the modified equation of state

$$m_\alpha = \frac{\mathscr{M}_\alpha PV}{RT}$$

where $\mathscr{M}_\alpha$ is the mean molar mass of air, calculated on the assumption that air is 77% $N_2$ and 23% $O_2$ by mass. $P$ is the atmospheric pressure in the laboratory, $T$ the laboratory temperature (K), and $V$ the volume of the flask.

The mass of the truly empty flask is found by subtracting the mass of included air from the combined mass measured on the balance.

When the flask is weighed after it has cooled, it contains condensed liquid, vapor of the volatile liquid, and air. The mass of air in the flask at this stage is found in the same way as before, except that $P$ is now the partial pressure of air in the flask. The partial pressure is the measured atmospheric pressure less the vapor pressure of the unknown in the flask. Your instructor will post a table of the vapor pressures of the various unknowns in the temperature range of interest. The total mass of unknown (liquid and vapor) is found by subtracting the mass of the truly empty flask and the mass of the included air from the combined mass measured on the balance.

## PROCEDURE

### SAFETY NOTES
Some of the volatile liquids used in this experiment are flammable. Keep open flames away from all of them.

Do not handle the hot flasks with your fingers during the quenching step

The experiment is best carried out by students working in pairs. You may collaborate in recording data, but should work independently on the prelaboratory and report sheets. Do two determinations on the same unknown.

### 1. Preparing a Dumas Bulb

If specially constructed Dumas bulbs are available, use them and follow any modifications of the procedure suggested by your instructor. Otherwise, prepare two bulbs as follows.

Thoroughly clean two Erlenmeyer flasks (200–250 mL) by washing them in hot water and detergent. Rinse each flask with distilled water, dry off the outsides with a clean towel, and invert the flasks on a towel to drain dry. If water droplets adhere to the insides at this stage, repeat the washing and rinsing process properly. Carry out a final drying of the flasks by passing them over a gentle blue burner flame, holding the flask by the neck with a folded strip of paper. Move the flask continuously over the flame so as to avoid local overheating. Place the flask base down on a clean, dry towel and let it cool to room temperature. Make sure that no water vapor condenses in the flask at this stage.

For each flask cut a 5-cm square of aluminum foil and use it to cover the mouth of the flask. Secure the foil tightly in place with a rubber band and trim off any excess foil with scissors. Leave as much of the neck of the flask exposed as possible. Turn any small amount of foil back over the rubber band. The flask will later be immersed up to the level of the foil in water, so be sure that no ragged edges remain which might pick up water.

Wipe the flask and foil with a barely damp cloth so as to remove any fingerprints and to discharge any static electricity. Let the flask dry for at least 10 minutes and then weigh the assembly to the nearest milligram on the fine balance. Record the mass and keep handling of the flask to a minimum after this point.

## 2. Introducing the Sample

Your instructor will assign you a specific unknown and tell you its empirical formula and boiling point. From the labeled bottle corresponding to your unknown, draw about 5 mL in the syringe provided. Use only the syringe corresponding to your bottle of unknown so as to avoid contamination. Wipe the needle dry and inject the liquid into the flask through the foil cover. Keep the size of the injection hole as small as possible and use a slow steady pressure on the syringe. Take care to avoid splashing the liquid onto the underside of the foil or the top of the neck of the flask. Wait until the last drop of liquid has been discharged and carefully withdraw the syringe, taking care not to enlarge the hole in the foil, which must be as small as possible if good results are to be obtained. Wipe off any liquid accidentally spilled on the foil top with a piece of filter paper.

## 3. Evaporating the Sample

Set up a water bath by half filling a 1-L beaker with distilled water. (Tap water may deposit a scum on the flask if the water is hard. See Section 11.6 of *Chemical Principles*.) Set up a ring stand with wire gauze and burner and set the beaker on the ring. With a ring stand clamp, firmly secure the neck of the flask and immerse the flask in the water bath so that as much as possible of the flask is inside the bath. The clamp should be aligned with and touch the lip of the beaker and the neck of the flask should be secured so that as much as possible of the flask will be under water when the bath is filled. Make sure the flask is not in contact with the sides of the beaker, so that there is no obstruction to the circulation of water in the bath. Top up the water level so that it is as close to the top as possible, but take care not to let the water level quite reach the foil cap. Have your instructor inspect your apparatus before proceeding.

Now, slowly and steadily heat the water bath to a temperature 10°C above the boiling point of your unknown, but in no case above 95°C. Once this temperature is reached, maintain it by careful manipulation of the burner, adding heat as required and stirring the water in the bath continuously to keep it circulating. Take care not to splash water onto the foil cap in this step.

It is of the greatest importance that the temperature not drop at any stage during the evaporation of the unknown liquid as air may be drawn into the flask to compensate for the drop in pressure and there may not be enough vapor to completely expel the air. Plan a careful and slow rise in the bath temperature, at the same time not letting the bath water boil.

The sample should have completely evaporated within 10 minutes. You can check this by watching for the disappearance of the Schlieren pattern above the pinhole. A Schlieren pattern results from the disturbance in the refractive index of the air caused by the presence of organic vapors. You may have noticed this effect above the gas cap on your car when you fill up on a hot day, or may have seen the shimmering of the air above a hot outdoor fire. At your instructor's discretion you may wish to add an additional 1 mL of sample and continue heating at the constant temperature for another 10 minutes or so, so as to be sure that all the air has been driven from the flask. However, if you have followed these instructions well, then this second addition should not normally be needed.

Once the vapor expulsion has ceased, hold the bath temperature constant for another minute. Record this final temperature, and at once remove the flask from the water bath by loosening the clamp at the stand and sliding the clamp up the stand. Quench the vapor by dipping the flask into another 1-L beaker half filled with cold water. After a few minutes, remove the flask from the beaker, dry off the outside and set the flask aside on a clean dry towel to equilibrate to room temperature. When equilibration is complete and the outside of the flask is completely clean and dry, reweigh the assembly to the nearest milligram on the same balance as before.

### 4. Recording Barometric Pressure and Laboratory Temperature

Read and record the barometric pressure registered on the laboratory barometer. Correct this pressure to standard temperature and gravity, according to the instructions posted by the barometer. Your instructor will provide you with the latitude of the laboratory.

Record the temperature in the laboratory at the place where the flask equilibrated to room temperature. Be alert for spurious readings caused by burner flames in the neighborhood. It is always wise to extinguish the burner when it is no longer needed.

### 5. Determining the Volume of the Flask

Remove but retain the foil cap and rubber band from the flask. Discard the liquid sample in the marked disposal jar and rinse the flask several times with tap water. Carefully dry off the outside of the flask and fill it almost to the rim with water. Take the flask, foil, and rubber band, place them on a rough balance, and carefully top up the flask with distilled water from your washbottle. When the flask is completely filled to the rim with water, weigh the assembly as accurately as you can. (The total mass is outside the range of the fine balance.) Subtract the mass of the empty flask, foil, and band assembly and find the volume of the flask by noting that the density of water at 22°C is 0.998 g/mL.

Name _____  Instructor _____

Partner _____

## DATA SHEET

Unknown identifier: _____  Boiling point: _____ °C _____ K

Empirical formula of unknown: _____

|  | Run 1 | Run 2 |
|---|---|---|
| Mass of flask assembly empty | 71.975 g | _____ |
| Mass of flask assembly + cooled sample | 72.479 g | _____ |
| Mass of flask assembly + water | 223.559 g | _____ |
| Mass of water | 151.584 g | _____ |
| Volume of water | 151.889 cm³ | _____ |
| Final bath temperature | 90 °C | _____ °C |
|  | 363 K | _____ K |
| Room temperature | 27 °C | _____ °C |
|  | 300 K | _____ K |
| Barometric pressure reading | 764.4 mmHg | _____ |
| Corrections | _____ | _____ |
| Corrected pressure | _____ | _____ |
| Vapor pressure of unknown at room temperature | 102 mmHg | _____ |

$0.997992 \text{ g/cm}^3$   $102 \text{ mmHg} \times \dfrac{1 \text{ atm}}{760 \text{ mmHg}}$

$151.584 \text{ g} \times \dfrac{\text{cm}^3}{0.997992 \text{ g}} =$

$1.006 \text{ atm} - 0.134 \text{ atm}$

Name _____  Instructor _____

Partner _____

## DATA SHEET

Unknown identifier: _____  Boiling point: _____ °C _____ K

Empirical formula of unknown: _____

|  | Run 1 | Run 2 |
|---|---|---|
| Mass of flask assembly empty | _____ | _____ |
| Mass of flask assembly + cooled sample | _____ | _____ |
| Mass of flask assembly + water | _____ | _____ |
| Mass of water | _____ | _____ |
| Volume of water | _____ | _____ |
| Final bath temperature | _____ °C | _____ °C |
|  | _____ K | _____ K |
| Room temperature | _____ °C | _____ °C |
|  | _____ K | _____ K |
| Barometric pressure reading | _____ | _____ |
| Corrections | _____ | _____ |
| Corrected pressure | _____ | _____ |
| Vapor pressure of unknown at room temperature | _____ | _____ |

## WORKSHEET

Carry out all the calculations for Experiment 6 on this sheet, attach it to your report sheet, and submit the two together. Do your calculations directly on this sheet. Do not do them on scratch paper and then copy them on to this sheet. Make no erasures and delete incorrect entries with a single line so as to leave the original data legible. Use a pen.

$$\frac{(28.8 \text{ g/mol})(1.006 \text{ atm})(0.1516 \text{ L})}{(0.0821 \text{ L·atm/mol·K})(300 \text{ K})}$$

$$\frac{\text{mol}}{24.63 \text{ L·atm}} \cdot \frac{4.392 \text{ g·atm·L}}{\text{mol}}$$

$$\frac{28.8 \text{ g/mol}}{(0.0821)(300)} (0.872)(0.1516) = 0.155 \text{ g}$$

147

# EXPERIMENT 8

# Evaluation of Avogadro's Number from the Measured Density and X-Ray Diffraction Data for a Crystalline Organic Compound

### AIM

To determine Avogadro's number from given X-ray diffraction measurements and an experimentally measured density and to analyze the uncertainties in this determination.

### EQUIPMENT

From stockroom: 5-mL pipet
From locker: 30-mL beakers
Test tube
Laboratory: Fine balance

### REAGENTS

Crystalline samples of various organic compounds
Saturated aqueous solution of potassium iodide
Distilled water

## BACKGROUND

### Associated Readings

The following sections of *Chemical Principles* should be read in conjunction with Experiment 8.

**Chapter 10** States of Matter: Condensed States
              10.3 The Crystalline State
              10.4 Unit Cells and Lattices
              10.6 X-ray Crystallography

The structure of a crystal may be thought of as made up by the regular translational repetition in three dimensions of some basic unit of structure. This unit of structure, in turn, may be viewed as enclosed in a parallelepiped whose three edge vectors define both the repeat distances in the crystal and their orientations in space. This parallelepiped is known as the **unit cell**. Its shape and size are defined in terms of three **axial lengths**, $a$, $b$, and $c$, and three **interaxial angles**, $\alpha = \widehat{bc}$, $\beta = \widehat{ac}$, and $\gamma = \widehat{ab}$.

The exact shape of the unit cell in a particular case depends on the overall symmetry of the crystal structure. Seven crystal systems based on crystal symmetry are recognized, and listed in Table 8.1. The presence of particular combinations of axes of symmetry in the crystal structure determines the crystal system, defines the values of one or more of the interaxial angles, and may lead to certain equalities in the lengths of the unit cell axes. An $n$-fold axis of symmetry is an operator that leads to a repetition of some object in space by rotation of the object through $2\pi/n°$. An unlabeled hexagonal pencil has a six-fold axis of rotation coincident with the graphite core. In crystals, $n$ is limited to 1, 2, 3, 4, or 6. An unlabeled cylindrical pencil, however, would have $n = \infty$.

The general expression for the volume of the unit cell is that for the generalized parallelepiped and is given by

$$V = abc(1 - \cos^2\alpha - \cos^2\beta - \cos^2\gamma + 2\cos\alpha \cos\beta \cos\gamma)^{1/2}$$

This expression is considerably simplified where two or more of the interaxial angles are exactly 90°.

The ratio of the mass of the unit cell contents to the cell volume is the density of the solid, an experimentally measurable quantity. The mass of the unit cell contents is, of course, the sum of the masses (in grams) of the atoms it contains and is given by:

$$\text{Mass} = (\Sigma \text{ Atomic masses of atoms in cell})/N_A$$

where $N_A$ is Avogadro's number.

We thus have the relationship:

$$\text{Density} = \rho = \text{Mass/Volume} = (\Sigma \text{ Atomic masses in cell})/N_A V.$$

**TABLE 8.1** The Seven Crystal Systems

| Crystal System | Parameters |
| --- | --- |
| Triclinic | $a \neq b \neq c;\ \alpha \neq \beta \neq \gamma$ |
| Monoclinic | $a \neq b \neq c;\ \alpha = \gamma = 90°;\ \neq \beta$ |
| Orthorhombic | $a \neq b \neq c;\ \alpha = \beta = \gamma = 90°$ |
| Tetragonal | $a = b \neq c;\ \alpha = \beta = \gamma = 90°$ |
| Rhombohedral | $a = b = c;\ \alpha = \beta = \gamma\ (\neq 90°)$ |
| Cubic | $a = b = c;\ \alpha = \beta = \gamma = 90°$ |
| Hexagonal | $a = b \neq c;\ \alpha = \beta = 90°;\ \gamma = 120°$ |

# Evaluation of Avogadro's Number from the Measured Density of X-Ray Diffraction Data for a Crystalline Organic Compound

Where crystals of a solid are available, the volume of the unit cell can be determined from X-ray diffraction measurements and, in most cases, so can the nature of the unit cell contents. If the density of the crystals is then carefully measured, we have an estimate of Avogadro's number:

$$N_A = (\Sigma \text{ Atomic masses})/\rho \cdot V$$

In this experiment we determine the density of a crystalline organic compound and from data on the unit cell dimensions and contents establish the unit cell volume. Armed with this information we evaluate Avogadro's number and estimate its uncertainty.

## X-Ray Diffraction

Diffraction occurs whenever radiation of a particular wavelength is incident upon an object or series of objects whose dimensions are of the same order of magnitude as the wavelength.

X rays, discovered in the late 19th century by Wilhelm Roentgen, were shown to have wave character when Max von Laue showed that they were diffracted by crystals. X-ray wavelengths lie in the nanometer range (about $10^{-10}$ m) and are comparable in magnitude to the distances between atoms in molecules and between molecules in crystals. Because of their three-dimensionally periodic structure, crystals act as three-dimensional diffraction gratings for X rays. The study of the diffraction behavior of crystals and X rays is known as **X-ray crystallography.** Most of our knowledge of the three-dimensional structure of molecules comes from the study of X-ray diffraction patterns. X-ray data were used in the structure determination of DNA and of the proteins myoglobin and hemoglobin in the late 1950's. The method is used routinely today as a powerful analytical tool for the determination of molecular structure.

The geometrical conditions for diffraction by a crystal are given by **Bragg's Law,** which states that:

$$n\lambda = 2d \sin\theta$$

where $n$ is an integer, $\lambda$ is the wavelength of the X rays, $\theta$ is the angle of incidence of the direct beam, equal also to the angle of reflection, and $d$ is the perpendicular separation between members of a particular family of parallel planes in the crystal.

Crystallographers find the lengths of the unit cell axes and interaxial angles for a particular crystal by calculating $d$-values for a sufficiently large number of reflections from different planes in the crystal. The techniques for doing this lie outside the scope of this course but you will likely encounter them if you go on to take a course in physical

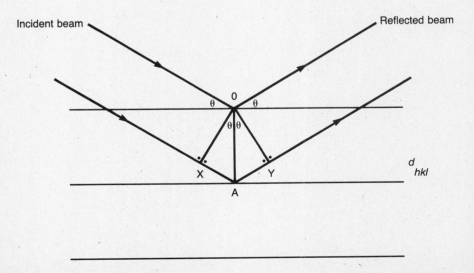

**Figure 8.1**
Bragg's law for reflection of X-rays. Constructive interference between incident and diffracted beams occurs only when the path difference between rays reflected from successive planes is an integral number of wavelengths. For the planes shown, the path difference is XA + AY or $2d_{(hkl)} \sin \theta$.

## TABLE 8.2  Unit Cell Data for Some Crystalline Organic Compounds*

| Compound | Formula | System | a | b | c | α | β | γ | Z |
|---|---|---|---|---|---|---|---|---|---|
| Biphenyl | $C_{12}H_{10}$ | Monoclinic | 812(1)# | 564(1) | 947(2) | 90.0(—) | 95.4(2) | 90.0(—) | 2 |
| Dibenzyl | $C_{14}H_{14}$ | Monoclinic | 1277(2) | 612(1) | 770(1) | 90.0(—) | 116.0(2) | 90.0(—) | 2 |
| Anthracene | $C_{14}H_{10}$ | Monoclinic | 856(1) | 604(1) | 1116(1) | 90.0(—) | 124.7(1) | 90.0(—) | 2 |
| Naphthalene | $C_{10}H_8$ | Monoclinic | 824(1) | 600(1) | 865(1) | 90.0(—) | 122.9(1) | 90.0(—) | 2 |
| Phenanthrene | $C_{14}H_{10}$ | Monoclinic | 846(1) | 616(1) | 947(1) | 90.0(—) | 97.7(2) | 90.0(—) | 2 |
| Fluorene | $C_{13}H_{10}$ | Orthorhombic | 849(1) | 572(8) | 1897(1) | 90.0(—) | 90.0(—) | 90.0(—) | 4 |
| Hexamethylbenzene | $C_{12}H_{18}$ | Triclinic | 892(1) | 886(1) | 530(1) | 44.5(1) | 116.7(1) | 119.6(1) | 1 |
| Anisic Acid | $C_8H_8O_3$ | Monoclinic | 1698(3) | 1095(2) | 398(1) | 90.0(—) | 98.71(1) | 90.0(—) | 4 |

*Data from "Structure Reports," a comprehensive series summarizing structural studies by diffraction methods. Axial lengths in pm, angles in degrees.

#For this experiment, 812(1) ≡ 812 ± 1.0 etc.
90.0(—) means exactly 90.0°

chemistry. Unit cell dimensions and crystal systems are known for many thousands of crystalline compounds, and values for a number of organic crystals used in this experiment are given in Table 8.2. Also included in the table are values of Z, the number of molecules present in the unit cell as determined by X-ray structure determination. This number must be taken into account in calculating the sum of atomic masses in the unit cell. Note that in all but the triclinic system this number is usually greater than 1.

X-ray measurements, like all scientific measurements, are subject to error. Estimates of the uncertainty in the various parameters are given in the table. These uncertainties are used to define the uncertainty in the unit cell volume and that uncertainty is combined with that in the density determination to set a limit to the accuracy of the value of $N_A$ found.

## PROCEDURE

### SAFETY NOTES
Be sure to use a pipet bulb when filling the 5-mL pipet. Do not pipet solutions by mouth.

Dispose of the potassium iodide solutions in the marked containers.

### 1. Calibration of the Pipet

Clean the pipet with cold water and detergent solution and rinse it with distilled water. Make sure that no water droplets adhere to the inside surface.

Determine the mass of a clean, dry 30-mL beaker to the nearest milligram. Pipet 5 mL of distilled water into the beaker and reweigh it to determine the mass of water transferred. Be careful *not* to force the last drop of water out of the pipet. By assuming the density of distilled water at 20°C to be 0.998 g/cm³, determine the actual volume of water delivered by the pipet.

### 2. Determination of Crystal Density

Add about 5 mL of the saturated potassium iodide solution (density ~ 1.6 g/cm³) and about 5 mL of distilled water to a clean test tube and shake it vigorously to achieve a homogeneous solution (density ~ 1.3 g/cm³).

To avoid problems arising from occluded air bubbles in the crystals, take a few crystals of your assigned compound and using a glass stirring rod with a rounded end gently grind them to a powder on a watchglass. Add the powder to the solution in the test tube, shake vigorously for about 30 seconds and allow the crystals to settle.

If the crystals have sunk to the bottom of the test tube, their density is greater than that of the solution. Increase the solution density by drop-by-drop addition of saturated KI solution, shaking the tube vigorously between additions then allowing the contents to settle, until the crystals neither sink nor float to the surface but remain suspended in the body of the solution. At this point the solution and crystal have the same density. It may be that if the densities of solid and solution are very closely matched, a few crystals may float to the surface, but none should remain on the bottom of the tube.

If, however, the crystals float to the surface when first added, the solution is too dense and should be diluted by dropwise addition of distilled water until the same result is achieved. This method of equalizing the density of solid and solution is capable of yielding very accurate results, and it is worth spending some time to adjust the density of the solution as accurately as possible.

When the solution and the crystal are deemed to have the same density, then pipet out 5 mL of the solution into a previously weighed clean, dry 30-mL beaker and weigh the beaker and its contents to the nearest milligram to establish the mass of solution transferred. Use the calibrated volume of the pipet to determine the solution density. Repeat the density determination on a separate sample of the material so as to obtain an estimate of the precision of your determination.

## 3. Treating the X-Ray Data

Use the data in Table 8.2 to calculate the volume of the unit cells. Estimate the uncertainty in the volume by calculating the "worst case" maximum and minimum volumes by assuming first that all the axial lengths are increased by the amount of the given uncertainty and second that all are decreased by the same amount. Check to see the effect on the volume of increasing or decreasing the angles in this way, and incorporate the appropriate values in the calculations. Note that in the monoclinic case where the general expression for the volume reduces to

$$V = abc \sin\beta$$

increasing the angle (all $\beta$ are obtuse by convention) decreases the volume. The triclinic case is more complicated and the effect of the uncertainty in each angle must be tested separately. Remember that angles of exactly 90° fixed by the crystal symmetry have no associated uncertainty.

Use the remaining data in Table 8.2 to find the sum of the atomic weights in the unit cell of your compound and combine your results to find $N_A$. Repeat the worst case procedure in combining your volume and density data to estimate the uncertainty in your determination. Note that your density is expressed as $g/cm^3$ whereas the unit cell volume has units of $(pm)^3$. Be sure to reduce these quantities to a common volume base (e.g. $m^3$).

Name _____  Instructor _____

## PRELABORATORY PREPARATION SHEET

1. What is meant by the expression "unit cell of a crystal"?

   _____

   _____

2. What is the relationship between the lengths of the unit cell axes in the tetragonal system?

   _____

3. What is the relationship between the unit cell angles in the monoclinic system?

   _____

4. Sucrose, $C_{12}H_{22}O_{11}$ crystallizes in a monoclinic unit cell with dimensions $a = 1100$, $b = 870$, $c = 770$ pm, $\beta = 103.5°$. Its measured density is 1.588 g/cm$^3$. Use the value of Avogadro's number given in *Chemical Principles* to calculate the number of sucrose molecules in the unit cell.

   _____

   _____

   _____

   _____

   _____

   _____

   _____

Name _____  Instructor _____

Partner _____

## DATA AND REPORT SHEET

                                                                  Run 1              Run 2

**1.** Calibration of 5-mL pipet:

    Mass of beaker and distilled water     _____     _____

    Mass of beaker empty     _____     _____

    Mass of water transferred     _____     _____

    Volume of water transferred     _____     _____

    Mean volume of pipet     _____ $\pm$ _____

**2.** Determination of density:

    Mass of beaker + solution     _____     _____

    Mass of beaker empty     _____     _____

    Mass of solution transferred     _____     _____

    Mean mass of solution     _____ $\pm$ _____

    Mean density     _____ $\pm$ _____

**3.** X-ray data:
Enter the unit cell data for your compound as given in Table 8.2. Specify units throughout.

    Compound _____

    $a =$ _____ $\pm$ _____

    $b =$ _____ $\pm$ _____

    $c =$ _____ $\pm$ _____

    $\alpha =$ _____ $\pm$ _____

    $\beta =$ _____ $\pm$ _____

    $\gamma =$ _____ $\pm$ _____

    Calculated unit cell volume     _____

    Calculated maximum unit cell volume     _____

    Calculated minimum unit cell volume     _____

    Mean unit cell volume taken     _____ $\pm$ _____

Name _____    Instructor _____

Partner _____

## DATA AND REPORT SHEET

                                            Run 1              Run 2

**1.** Calibration of 5-mL pipet:

    Mass of beaker and distilled water   _____      _____

    Mass of beaker empty   _____      _____

    Mass of water transferred   _____      _____

    Volume of water transferred   _____      _____

    Mean volume of pipet   _____ ± _____

**2.** Determination of density:

    Mass of beaker + solution   _____      _____

    Mass of beaker empty   _____      _____

    Mass of solution transferred   _____      _____

    Mean mass of solution   _____ ± _____

    Mean density   _____ ± _____

**3.** X-ray data:
Enter the unit cell data for your compound as given in Table 8.2. Specify units throughout.

    Compound _____

    $a =$ _____ ± _____

    $b =$ _____ ± _____

    $c =$ _____ ± _____

    $\alpha =$ _____ ± _____

    $\beta =$ _____ ± _____

    $\gamma =$ _____ ± _____

    Calculated unit cell volume _____

    Calculated maximum unit cell volume _____

    Calculated minimum unit cell volume _____

    Mean unit cell volume taken _____ ± _____

**4.** Atomic Mass Data (Assumed Error Free):

    Formula of compound _____

    Molar mass     _____

    $Z$     _____

    $Z \times$ molar mass     _____

**5.** Calculation of $N_A$:

    Calculated value for $N_A$     _____

    Maximum "worst case" value for $N_A$ _____

    Minimum "worst case" value for $N_A$ _____

    Value of $N_A$ taken _____ $\pm$ _____
The accepted value of $N_A$ is currently approximated as $6.022 \times 10^{23}$ mol$^{-1}$. Calculate the difference between your value and the accepted value.

_____

**6.** Offer a possible explanation if this difference lies outside your experimentally determined limits of error.

_____

_____

_____

_____

_____

_____

4. Atomic Mass Data (Assumed Error Free):

   Formula of compound _____

   Molar mass _____

   $Z$ _____

   $Z \times$ molar mass _____

5. Calculation of $N_A$:

   Calculated value for $N_A$ _____

   Maximum "worst case" value for $N_A$ _____

   Minimum "worst case" value for $N_A$ _____

   Value of $N_A$ taken _____ ± _____
   The accepted value of $N_A$ is currently approximated as $6.022 \times 10^{23}$ mol$^{-1}$.
   Calculate the difference between your value and the accepted value.

   _____

6. Offer a possible explanation if this difference lies outside your experimentally determined limits of error.

   _____

   _____

   _____

   _____

   _____

   _____

## WORKSHEET

Carry out all the calculations for Experiment 8 on this sheet, attach it to your report sheet, and submit the two together. Do your calculations directly on this sheet. Do not do them on scratch paper and then copy them onto this sheet. Make no erasures and delete incorrect entries with a single line so as to leave the original data legible. Use a pen.

# EXPERIMENT 9

# Water Analysis

### AIM

To determine certain chemical components in a water sample. Quantitative analyses will be carried out for dissolved oxygen and for water hardness. Acidity and dissolved precipitable anion concentration will be determined qualitatively.

### EQUIPMENT

| | |
|---|---|
| From stockroom: | 25-mL pipet |
| | 50-mL buret |
| From locker: | 500-mL glass bottle |
| | 250- and 300-mL Erlenmeyer flasks |
| | Stirring rod |
| | Test tube |
| | Medicine dropper |
| | Glass filter funnel |
| Laboratory: | Buret stand and clamp |
| | 100-mL graduated cylinder |
| | Rubber tubing |
| | Dispensing pipets |
| Other: | Student-supplied water container |

## 190 EXPERIMENT 9

## REAGENTS

Student-supplied water sample
Manganese(II) sulfate solution
Sodium hydroxide/potassium iodide solution
Reference chloride solutions
$0.025M$ sodium thiosulfate solution
Stock sulfuric acid solution
Standard calcium carbonate solution
Ethylenediaminetetracetic acid solution (EDTA)
$1M$ $AgNO_3$ solution, $1M$ $HNO_3$
Eriochrome black T indicator, pH paper
Concentrated sulfuric acid, $H_2SO_4$

## BACKGROUND

### Associated Readings

The following sections of *Chemical Principles* should be read in conjunction with Experiment 9.

**Chapter 11** Water
                11.1 Properties of Water
**Chapter 12** The Properties of Solutions
                12.3 Solubility (Solubility of Gases)

Water is essential for the support of all life forms and is a vital requirement in a technical civilization.

The chemical reactions that sustain life all take place in aqueous solution, with water itself frequently involved as a reactant, as in the process of photosynthesis by which green plants convert carbon dioxide and water into complex sugars, liberating oxygen as a by-product.

$$6CO_2(g) + 6H_2O \longrightarrow C_6H_{12}O_6(aq) + 6O_2(g)$$

Equally important are the processes in the lung whereby oxygen is transferred to hemoglobin for transport throughout the circulatory system, where water plays a vital role—dehydration of the lungs is a fatal condition.

Besides the normal requirements of each individual for drinking water to maintain the 70% water content of the body, agricultural and industrial processes require enormous amounts of water. Thus adequate rainfall or irrigation is needed for crop production; 600 kg of water are used in the preparation of 1 kg of fertilizer, 300 kg in the production of 1 kg of steel, and more than 200 kg in the production of a single copy of this laboratory manual from the original tree. A plentiful and reliable supply of water is thus essential to civilization, and the development of all societies has been closely linked to availability and control of water supply.

Although about 80% of the planet's surface is covered with water, surprisingly little of the total estimated mass of water, some $1.3 \times 10^{24}$ kg or about 5% of the mass of the earth, is available for ready use. Over 97% of that total, for example, is in the seas and oceans as a dilute saline solution not directly useful either as drinking water or for agricultural purposes. This huge volume of water, however, plays an important physical role, acting as a heat sink to absorb most of the tremendous amount of solar energy that falls on the earth daily. Liberation of this energy by radiation during hours

## Water Analysis

of darkness helps to stabilize the surface temperature of the earth. Without this water buffer the earth's surface temperature would fluctuate between $-150°C$ and $+120°C$ every 24 hours.

The ability of water to sustain life is closely linked to its ability to dissolve molecular oxygen from the atmosphere. Dissolved oxygen also results from photosynthetic processes taking place in aquatic plants in the oceans. Over three-quarters of all the oxygen produced by photosynthesis is in the oceans. It is this aspect of the ecocycle, more than any other, that prompts concern over the effects of oceanic pollution. By killing the plant life of the sea we cut off a major source of oxygen needed to sustain human life. The problem is a very real one and demands immediate action. We have traditionally used the oceans as dumping grounds for many of our wastes and the process continues at an alarming rate; there is already a very real prospect of the Mediterranean becoming a dead sea in our lifetime.

The minimum concentration of dissolved oxygen needed to support life is only 4 mg/L or four parts of dissolved oxygen per million parts of water (4 ppm). The equilibrium concentration of dissolved oxygen in water, although it varies with temperature, is normally higher than that: 14.6 ppm at 0°C, 11.3 ppm at 10°C, 9.2 ppm at 20°C, and 7.6 ppm at 30°C. It might be thought, therefore, that a sufficiency of dissolved oxygen would always be present. Unhappily, the process of solution is a very slow one and is ultimately dependent on the surface-to-volume ratio of any body of water. A raging glacial torrent in a mountain stream, where the water is constantly being broken up and tossed into the air, will quickly and naturally become saturated with oxygen; a large, still deep lake may only achieve total saturation over centuries, if at all.

Where excessive pollution occurs, the processes occurring in the water that absorb oxygen may rapidly outstrip the rate at which oxygen is produced by photosynthesis or absorbed from the atmosphere. Under these conditions all of the oxygen may be removed and **eutrophication,** a condition of total oxygen depletion, results. This condition is discussed in more detail in Section 11.6 of *Chemical Principles*.

Aerobic bacteria use oxygen to degrade complex organic molecules to carbon dioxide and water. The demand for oxygen of organisms in a sample of water is referred to as **biochemical oxygen demand,** or **BOD.** It is determined by measuring the mass of oxygen consumed in a sample of water of known volume in a fixed period of time. The water sample is first diluted with oxygen-saturated distilled water to make sure that an excess of oxygen is initially present. Its oxygen content is immediately measured and then measured again after five days. From the decrease in oxygen content, BOD is calculated as:

$$BOD = mg \text{ dissolved } O_2 \text{ consumed/L of water}$$

In this experiment we will quantitatively determine the dissolved oxygen content of your water sample by Winkler's method. The sample is first treated with excess manganese(II) sulfate solution then with an alkaline solution of potassium iodide. The following reactions are important:

$$Mn^{2+}(aq) + 2OH^-(aq) \longrightarrow Mn(OH)_2(s)$$

$$4Mn(OH)_2(s) + O_2(aq) + 2H_2O \longrightarrow 4Mn(OH)_3(s)$$

The $Mn(OH)_2$ initially formed reacts with the dissolved oxygen. The process is a heterogeneous reaction, involving combination of a gas with a colloidal solid, and is slow. (See Section 12.8 of *Chemical Principles*).

The amount of $Mn(OH)_3$ formed is determined by reaction with iodide ion, which is inert in basic solution, but in acidic solution reacts with $Mn(OH)_3$ to form $Mn^{2+}$ and iodine. The sample is now acidified with sulfuric acid and the following reactions take place:

$$Mn(OH)_2(s) + 2H^+(aq) \longrightarrow Mn^{2+}(aq) + 2H_2O$$

$$2Mn(OH)_3 + 2I^-(aq) + 6H^+(aq) \longrightarrow 2Mn^{2+}(aq) + I_2(aq) + 6H_2O$$

The iodine formed may be titrated against standard thiosulfate solution, using starch as an indicator, to determine the amount of iodine formed. From the various equations, the amount of dissolved oxygen is readily calculated.

$$I_2(aq) + 2S_2O_3^{2-}(aq) \longrightarrow 2I^-(aq) + S_4O_6^{2-}(aq)$$

Most salts are soluble to a greater or lesser extent in water, and natural water supplies contain varying amounts of dissolved cations and anions of various kinds reflecting the nature of the ground over which the water has passed or is stored. Small amounts of $Na^+$ and $K^+$ occur in fresh water and cause no serious problems. Trouble is caused in some applications by the presence of $Ca^{2+}$, $Mg^{2+}$, and $Fe^{3+}$ ions because of their tendency to precipitate as insoluble salts, particularly carbonates, phosphates, and hydroxides. When this precipitation takes place in pipes or boilers, flow is often critically reduced and dangerous situations may arise. This problem is particularly severe in limestone bearing areas.

More commonly these ions present a problem because they combine with the anionic component of many soaps and detergents to form insoluble precipitates—the familiar scum in the sink or washing machine. Soaps are sodium or potassium salts of aliphatic fatty acids. These binary salts are soluble, but the ternary calcium or magnesium salts are not.

Water with a high concentration of $Mg^{2+}$ and $Ca^{2+}$ is referred to as **hard water,** and the degree of hardness is linked to the concentrations of these two cations. Water with a low concentration of these ions is referred to as being **soft.**

The hardness of your water sample will be determined by titration of the alkaline earth cations with ethylenediaminetetracetic acid (EDTA), a complexing agent. The titration is monitored by using another complexing agent, eriochrome black T (EBT), as an indicator. EDTA is a more powerful complexing agent than EBT and displaces EBT from its calcium complex. Ca(EBT) is a magenta shade whereas EBT itself is a deep blue color; Ca(EDTA) is colorless. As the titration proceeds, added EDTA removes $Ca^{2+}$ from Ca(EBT), and when all the calcium has been attached to EDTA at the end point, only the blue color of EBT is present. The color change is progressive, with magenta giving way to blue, so that the end point may not be too obvious in an initial run.

EDTA is a powerful complexing agent for nearly all metal ions forming exclusively 1:1 complexes. Discrimination between different metal ions is achieved by careful control of the pH of the solution. $Ca^{2+}$ and $Mg^{2+}$ are bound in a tight complex only in strongly basic solution, optimally at pH 10. It is necessary to use an appropriate buffer solution to ensure the proper pH of the sample during the EDTA titration. In the actual experiment you will standardize the stock solution of EDTA against a $CaCO_3$ solution of known concentration and then use the EDTA to determine the unknown $Ca^{2+}$ and $Mg^{2+}$ content of your sample.

The anion content of the water sample will be tested qualitatively by addition of $AgNO_3$ solution to a small sample. Insoluble precipitates of AgCl or $Ag_3PO_4$ formed when an acidified water sample is tested in this way indicate the presence of these anions. Their concentration will be estimated qualitatively by comparison of the quantity of precipitate formed in your sample and in a range of standard references.

A certain amount of chloride ion is almost always present in domestic water supplies as the result of chlorination procedures adopted in the course of purification. Chloride ion is an essential nutrient and one danger of the current antisalt fad is that chloride sufficient to sustain the stomach's need to synthesize hydrochloric acid may not be ingested. Excessive amounts of chloride or phosphate are likely to be the result of industrial or agricultural pollution of the water. Hydrochloric acid is used extensively in the production of steel, and wastes are commonly discharged into many lakes and rivers. Chloride ion also results from discharge of wastes by paper and pulp mills which use chlorine as a bleaching agent. Heavy rainfall, wind erosion, or poor plowing techniques may lead to run off of fertilizers from fields and produce a high phosphate content in water. Phosphate contamination was at one time a much more serious pollution problem when detergent manufacturers used complex phosphates in their products to solubilize calcium ion. The resultant increase in phosphate content of water supplies led to a prolific growth of certain plants and algae, which outstripped the ability of the water to absorb oxygen. In the resultant competition for dissolved oxygen, algae were much more successful than fish.

Natural water supplies in contact with air also absorb carbon dioxide, with production of the weak acid carbonic acid. This acid, found only in aqueous solution, interacts with water to form hydrated protons and bicarbonate anions:

$$CO_2(g) + H_2O \rightleftharpoons H_2CO_3(aq)$$

$$H_2CO_3(aq) \rightleftharpoons H^+(aq) + HCO_3^-(aq) \quad K_a = 4.3 \times 10^{-7}$$

Most natural water sources are therefore slightly acidic from this cause. To sustain life water must remain fairly close to neutral in character. A gradual control is being gained over the direct discharge of strongly acid or basic waste effluents into rivers and lakes, though these controls are often evaded by the socially irresponsible or through accident. A serious problem is arising in certain parts of the world with the steady increase in so-called acid rain, which is leading to a gradual increase in the acidity of many natural waters and the destruction of many plant species in affected areas. Ironically, this situation has arisen in part from attempts to solve local pollution problems. The construction of extremely tall chimneys to discharge sulfur dioxide-containing gases, and the removal of alkaline fly ash from these gases has led to a cleaner Cleveland but a less healthy Maine.

You test the pH of your water sample by a universal indicator paper, i.e., a paper designed to produce different colors for different pH throughout the range 1–14.

## PROCEDURE

Your water sample, secured in advance of this laboratory period, should be stored in a suitable clean glass or plastic container. A total volume of around 1 liter should suffice. Consider that the dissolved oxygen content of your sample will be affected by such factors as the conditions of storage and the aerobic organism content. You are encouraged to be adventurous in your choice of sample, but ordinary tap water will do.

### 1. Determination of Acidity

Clean and dry a glass stirring rod and use it to remove a drop of water. Touch the drop of water to a 3-cm length of universal pH paper. Match the color produced to the pH scale provided and record your observation.

There are two points to bear in mind for the future concerning this simple procedure:

**a.** Always touch the drop of solution to the paper, never dip the paper in the solution.
**b.** A 3-cm piece of paper is all that is required.

### 2. Determination of Precipitable Anion Content

Thoroughly clean and dry a test tube. Add 5 mL of your sample to the test tube (about 2 cm of water in the tube) and add 3 drops of $1M$ $HNO_3$. Shake the tube to mix the acid and water. If your sample was excessively basic, check the pH at this stage to be sure that the contents of the test tube are acidic. Add 3 drops of $1M$ $AgNO_3$ and again shake the tube gently. Note any precipitation and record the cloudiness of the solution as: none, faint, light, medium, or heavy.

Select the reference chloride solution which you think best matches your sample and repeat the above test on 5 mL of that solution. Repeat tests on the reference solutions until you find a match or until your sample produces results lying between two of the references. Record the chloride content.

### 3. Determination of Dissolved Oxygen

Take a glass filter funnel and attach to its stem a piece of rubber tubing about 25 cm in length. Introduce the tube into the clean, dry 500-mL glass bottle, as shown in Figure 9.1, and gently pour water from your sample into the bottle. Avoid any agitation of the system so as to minimize the amount of oxygen entering the water from the air during the transfer. Allow the bottle to overflow, remove the tubing, and place the stopper in the bottle. Water should overflow as the stopper is inserted. Make sure no air is trapped between stopper and water.

Take a long glass dispensing pipet, remove the stopper from the bottle, and introduce the pipet into the bottle so that the narrow tip rests on the bottom of the bottle. With a medicine dropper, drip 1 mL (20 drops) of the $MnSO_4$ solution into the pipet mouth. Gently raise the pipet from the bottle, allowing the $MnSO_4$ to drain into the body of the sample. With the same pipet, but with a different dropper, follow the same procedure to add 1 mL of the KI/NaOH solution to the bottle. Withdraw and discard the pipet, top up the bottle with additional water if needed, and restopper it, again taking care that no air is trapped.

Grasp the bottle firmly with a good grip on the stopper and invert the bottle several times to ensure complete reaction. Allow the bottle to stand until the precipitate has settled. (It is a good idea to proceed to some other determination while this is taking place.)

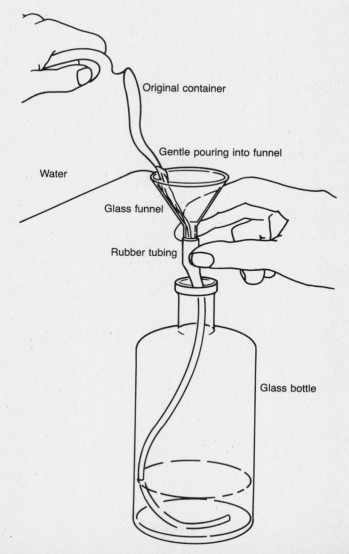

**Figure 9.1**
Transferring water to the glass bottle so as to avoid aeration.

With a dispensing pipet (*not* the same one used earlier) and following the same technique used before, add about 5 mL of concentrated sulfuric acid to the sample. Remove the pipet, replace the stopper, and once again being sure that there is no trapped air bubble, shake the bottle vigorously to dissolve the precipitate. This may take several minutes, so do not become impatient. If the precipitate does not completely dissolve after a few minutes of shaking, repeat the addition of sulfuric acid in the same way as before.

Using a graduated cylinder, measure 200 mL of the acidified sample into a 300-mL flask. Set up a 50-mL buret in a buret stand and ready it with the $0.25M$ thiosulfate solution. Titrate the 200 mL of sample with the thiosulfate until the color becomes a pale yellow. Add 5 mL of the starch solution to the flask. The deep blue color of the starch iodine complex should develop immediately. Continue the titration until the blue color just disappears and record the volume of thiosulfate used. Repeat the titration with a second 200-mL sample only if problems are encountered in the first titration.

## 4. Determination of Water Hardness

### a. Standardization of the EDTA Solution

In a clean, dry beaker collect about 75 mL of the standard (1.00 mg/mL) $CaCO_3$ solution. Transfer about 10 mL of this solution into a clean, dry 30-mL beaker and use it to rinse the 25-mL pipet. Now pipet exactly 25 mL of the original $CaCO_3$ solution into a clean, dry Erlenmeyer flask. Add two drops of the EBT indicator and 7 mL of the buffer solution. Swirl the contents of the flask to mix the reactants. With your wash bottle (distilled water only!) rinse down the sides of the flask to ensure that all reactants are in the body of the flask.

Carefully rinse the buret used in the last step first with distilled water and then twice with small amounts of the EDTA solution. Be sure to allow the EDTA solution to drain through the buret tip. Ready the buret with EDTA and titrate the $CaCO_3$ sample until the color changes from magenta to blue. This color change is very gradual; titrate to a distinct blue color and *keep the flask and its contents as a reference for subsequent titrations with EBT* in this step. Record the volume of EDTA required and repeat the titration for a second sample of $CaCO_3$. The two titrations should agree to within 0.5 mL.

### b. EDTA Titration of the Water Sample

Using a graduated cylinder, measure 100 mL of your water sample into an Erlenmeyer flask. Add two drops of EBT and 7 mL of buffer solution. Mix well and use a drop of the mixture to check that the pH is not less than 9. Add more buffer if needed to achieve a suitable pH. Titrate against EDTA to the same blue end point as before and record the volume of EDTA used. Carry out a duplicate titration if time permits.

Name _____  Instructor _____

## PRELABORATORY PREPARATION SHEET

1. The principal prelaboratory preparation for this experiment, besides reading carefully through the background material, is the collection of a suitable water sample. You will need about 1 liter of water and it should be stored in a clean glass or plastic container to obtain reliable indications on the dissolved oxygen content of the sample as nearly as possible at room temperature. Imaginative water sampling is encouraged. There are no "official" correct results in the usual sense for this experiment. Some obvious types of sample may suggest themselves after you have read the background material. Unless you are deliberately testing for low dissolved oxygen content, samples likely to contain significant amounts of microorganisms should not be stored for too long a period.

2. In the procedure for testing for dissolved oxygen content, you are repeatedly warned against trapped air bubbles in the sample. Why do you suppose that is?

3. In the EDTA titrations, the solution is buffered to high pH. Why?

4. Do some background research to find the structural formula for EDTA and the possible structural formula of its Ca complex. Give them here.

5. Do some background research into water purification for drinking purposes. Give a concise outline of the steps in a typical process with their effects. Feel free to use chemical equations and any additional space you may require.

197

_____

_____

_____

_____

_____

_____

6. If you have not previously used burets or pipets, read the Appendix—The Design, Use, and Care of Volumetric Glassware, for advice on how to use that equipment.

Name _____  Instructor _____

## DATA SHEET

1. Observed pH of sample: _____

2. Temperature of sample: _____

3. Appearance of sample: _____
   _____

4. Observed precipitable anion content:

   Character of precipitate: _____

   Closest reference concentration: _____

5. Thiosulfate titrations:

   |  | I | II |
   |---|---|---|
   | Volume of water sample: | _____ | |
   | Final buret reading | _____ | _____ |
   | Initial reading | _____ | _____ |
   | Volume $S_2O_3^{2-}$ used | _____ | _____ |

6. EDTA standardization:

   |  | I | II |
   |---|---|---|
   | Volume of $CaCO_3$ sample: | _____ | |
   | Final buret reading | _____ | _____ |
   | Initial reading | _____ | _____ |
   | Volume EDTA used | _____ | _____ |

7. Hardness determination:

   |  | I | II |
   |---|---|---|
   | Volume of water sample: | _____ | |
   | Final buret reading | _____ | _____ |
   | Initial reading | _____ | _____ |
   | Volume EDTA used | _____ | _____ |

Name _____  Instructor _____

## DATA SHEET

1. Observed pH of sample: _____

2. Temperature of sample: _____

3. Appearance of sample: _____
   _____

4. Observed precipitable anion content:

   Character of precipitate: _____

   Closest reference concentration: _____
5. Thiosulfate titrations:

|  | I | II |
|---|---|---|
| Volume of water sample: | | |
| Final buret reading | _____ | _____ |
| Initial reading | _____ | _____ |
| Volume $S_2O_3^{2-}$ used | _____ | _____ |

6. EDTA standardization:

|  | I | II |
|---|---|---|
| Volume of $CaCO_3$ sample: | | |
| Final buret reading | _____ | _____ |
| Initial reading | _____ | _____ |
| Volume EDTA used | _____ | _____ |

7. Hardness determination:

|  | I | II |
|---|---|---|
| Volume of water sample: | | |
| Final buret reading | _____ | _____ |
| Initial reading | _____ | _____ |
| Volume EDTA used | _____ | _____ |

Name _____  Instructor _____

# REPORT SHEET

1. Give the source of your water sample and the date collected.

   _____

2. Indicate any special features of the sample that might affect the analysis.

   _____

   _____

3. Give water temperature at time of analysis (from data sheet). _____
4. Note any odor, color, or suspension at the time of analysis.

   _____

   _____

5. pH of sample (from data sheet): _____

6. Precipitable anion content (ppm): _____
7. Calculation of dissolved $O_2$ content:

   a. Mean volume of $0.25M$ $S_2O_3^{2-}$ used: _____

   b. Calculate moles $S_2O_3^{2-}$ used: _____

   c. Equivalent moles $I_2$: _____

   d. Equivalent moles $Mn(OH)_3$: _____

   e. Equivalent moles $O_2$: _____

   f. Mass of $O_2$ in mg: _____
   g. Express dissolved oxygen content of your total sample as parts per million by weight.

   _____

8. Calculation of Hardness:
   The standard $CaCO_3$ solution has a concentration of 1 mg $CaCO_3$/mL.
   a. From the mean volume of EDTA required in the standardization calculate the volume of EDTA equivalent to 1 mg $CaCO_3$/mL.

   _____

   b. Give the mean volume of EDTA required in the hardness titration. _____

   c. To how many mg $CaCO_3$/mL does this correspond? _____

   d. Express the hardness of your sample as mg $CaCO_3$/L. _____

## WORKSHEET

Carry out all the calculations for Experiment 9 on this sheet, attach it to your report sheet, and submit the two together. Do your calculations directly on this sheet. Do not do them on scratch paper and then copy them on to this sheet. Make no erasures and delete incorrect entries with a single line so as to leave the original data legible. Use a pen.

# EXPERIMENT 10

# Ionic Interactions and Qualitative Analysis

## AIM

To identify each of nine unlabeled colorless ionic solutions by study of their interactions with one another.

## EQUIPMENT

From stockroom:  9 test tubes, each containing an unidentified ionic solution.
1-L beaker.
(Clean, dried and numbered test tubes are to be submitted to the stockroom in the 1-L beaker one week in advance of the experiment.)
From locker:  Test tube or
Watchglass and droppers
Laboratory:  Test tube rack

## REAGENTS

The unidentified ionic solutions are:

| | | |
|---|---|---|
| 1$M$ NaBr | 0.3$M$ Na$_2$SO$_4$ | 3$M$ NH$_4$Cl |
| 6$M$ NaOH | 0.5$M$ KClO$_3$ | 0.2$M$ BaCl$_2$ |
| 0.5$M$ Na$_2$S | 0.2$M$ KIO$_3$ | 0.5$M$ Pb(NO$_3$)$_2$ |

## BACKGROUND

### Associated Readings

The following sections of *Chemical Principles* should be read in conjunction with this Experiment 10.

**Chapter 2** The Language of Chemistry
   2.5 Important Types of Chemical Reactions
   2.6 States of Substances
   2.7 Net Ionic Reactions
**Chapter 12** The Property of Solutions
   12.3 Solubility (Solubility of Ionic Compounds in Water)
**Chapter 14** Chemical Equilibrium
   14.5 Solutions of Sparingly Soluble Substances: The Solubility Product (Precipitation Reactions)

Almost all salts exist in aqueous solution as discrete hydrated ionic species. When solutions of different salts are mixed, the ionic species present may remain in solution as independent entities, may combine to form a precipitate of an insoluble or sparingly soluble salt, or may react chemically to produce different species such as gases or insoluble precipitates that slowly hydrolyze in water. In this experiment you are provided with nine different salt solutions each numbered but otherwise unidentified. All the solutions are colorless. The object of the experiment is to identify each solution, using only the interactions, if any, occurring when small quantities of the solutions are mixed with one another. This type of experiment is often referred to as a nine-solution problem.

To carry out the experiment successfully, you must obviously know what interactions are to be expected for each mixture. Intelligent preliminary study is therefore essential, and as part of the prelaboratory preparation you will work out all the expected results of combining the solutions to yield binary mixtures. When you come to interpret your experimental data you will compare your observed results for the unknowns with those predicted for the given list of salts and on that basis decide the identity of the nine unknowns.

Although the experiment may be conducted unimaginatively by creating all possible binary combinations and noting the results, intelligent study of the expected results and some thought should allow you to be more selective and so reduce the number of mixings needed for positive identification. You may also wish to consider some ternary combinations; a number of these are useful for confirming an identification. A section is provided in the prelaboratory preparation for you to outline any scheme to speed the identification, and significant credit will be awarded for any good ideas there.

The following information is all that is strictly necessary to let you solve the problem, but you may consult the textbook (see the associated readings box) or any other reference work for additional information. In the case of sparingly soluble materials, you may wish to find out their solubility products, for example.

## Solubility of Salts

A salt is classified as soluble if its solubility is greater than 1 g/100 mL, as insoluble when its solubility is less than 0.1 g/100 mL, and as sparingly soluble when its solubility falls within these two limits. The solubility of most salts is affected by temperature. A salt that is sparingly soluble at room temperature may become soluble at higher temperatures; some salts that are soluble at room temperature become sparingly soluble when the temperature is raised.

For the problem at hand we may note the following:

1. All binary salts of $Na^+$, $K^+$, and $NH_4^+$, are soluble
2. All nitrates are soluble, as are chlorates and iodates, except for $Ba(IO_3)_2$ and $Pb(IO_3)_2$.
3. All carbonates are insoluble, except those of $Na^+$, $K^+$, and $NH_4^+$.
4. All sulfates are soluble, except those of $Ba^{2+}$, and $Pb^{2+}$, which are insoluble, and those of $Ca^{2+}$, $Ag^+$, and $Hg_2^{2+}$, which are sparingly soluble.
5. All chlorides and bromides are soluble, except AgCl, $Hg_2Cl_2$, AgBr, $Hg_2Br_2$, $PbBr_2$, and $HgBr_2$, which are insoluble. $PbCl_2$ is sparingly soluble at room temperature, soluble in hot water
6. All hydroxides are insoluble, except for those of Group IA, which are soluble, and those of $Ba^{2+}$ and $Ca^{2+}$, which are sparingly soluble.
7. Sulfides of Group IA are soluble, with noticeable hydrolysis to $H_2S$ taking place. Those of Group IIA and $Al^{3+}$ will form in precipitation reactions and hydrolyze only slowly. Other sulfides are insoluble and do not hydrolyze.

## Chemical Reactions

Two gases, recognizable by their characteristic odors, may be produced on mixing the solutions.

1. Ammonia, $NH_3$, is produced when solutions of ammonium salts react with strong base:

$$NH_4^+(aq) + OH^-(aq) \rightarrow NH_3(g) + H_2O.$$

2. Hydrogen sulfide, $H_2S$, is generated by addition of acid to sulfide solutions:

$$S^{2-}(aq) + 2H^+(aq) \rightarrow H_2S(g)$$

Recall that solutions of ammonium salts are acidic owing to the reaction:

$$NH_4^+(aq) + H_2O \rightarrow NH_3(aq) + H_3O^+(aq)$$

3. The Pb(II) ion may react with excess base to form the soluble complex ion $[Pb(OH)_3]^-$. An otherwise insoluble lead salt, such as lead sulfate, may be brought into solution in this way by addition of excess hydroxide ion. The ability of Pb(II) to form soluble complexes may produce some anomalous results when some solutions are mixed. Remember that whereas most precipitates form instantaneously on mixing, solubilization is much slower.
4. Only one oxidation-reduction reaction of importance is likely: the oxidation of sulfide ion to elemental sulfur by the action of iodate.

$$IO_3^-(aq) + 3S^{2-}(aq) \rightarrow I^-(aq) + 6OH^-(aq) + 3S(s)$$

Sulfide solutions are sometimes oxidized by air and may show contamination with small amounts of sulfite, sulfate, or sulfur.

## PROCEDURE

### SAFETY NOTES

Remember that $H_2S$ is a very toxic gas. It is, indeed, more toxic at a given concentration than HCN. Thus remember the rules for sampling odor and do not directly inhale either this gas or ammonia.

To avoid contamination of waste water by lead, temporarily store all mixings and rinsings in a beaker. At the end of the experiment, when the instructor has confirmed your identifications, the contents of the beaker and the lead solution should be placed in the marked disposal container.

One week before the scheduled period for this experiment, thoroughly clean nine test tubes with hot water and detergent. Make sure that all markings are removed from the test tubes. Carefully rinse the test tubes and set them upside down to dry in a test tube rack. Clean and rinse your 1-L beaker and dry it completely. Number your test tubes legibly, 1 through 9, place them open end up in the beaker and take the complete set to the stockroom. You will not necessarily receive the same set of test tubes back from the stockroom, so make sure that the numbering is carefully done. The stockroom staff reserve the right to reject test tubes improperly cleaned or numbered. A record is kept of all submissions.

On the day of the experiment collect a set of nine filled test tubes in a beaker from the stockroom. Carefully check the beaker and tubes for chips or cracks and immediately inform the stockroom staff if you find any. Included with the test tubes is a numbered slip identifying the particular set of unknowns you have received and having spaces for you to record your identifications. Put your name on the slip and hand it to your instructor for grading when you have completed the identifications.

Each test tube contains about 20 mL of solution—easily enough for all the mixings needed. Keep the quantity of solution used in each mixing to a minimum. (The record at the University of Virginia is a return of all test tubes with only 24 drops used, in just under six minutes.) You may obtain a refill of any one solution without penalty, but deductions will be made for additional refills. If you use droppers for mixing, be sure to rinse them thoroughly between each use so as to avoid contamination of one solution by another.

After your identifications have been checked, place any residual lead solution, together with the temporarily stored mixings and rinsings, in the marked disposal container. Wash all the other solutions down the sink with lots of cold water. Clean and rinse the test tubes and beaker and place them in your locker.

## WORKSHEET

Use this sheet to record any additional observations for Experiment 10. Submit it together with your identifications.

# EXPERIMENT 11

# Molar Mass Determination by Freezing Point Depression

### AIM

To establish the freezing point depression constant for cyclohexane and to use it to determine the unknown molar mass of a compound in cyclohexane solution.

### EQUIPMENT

From stockroom: Freezing point apparatus with narrow-range thermometer
From locker: 30-mL, 100-mL and 1-L beakers
25-mL graduated cylinder
Laboratory: Stand and clamp

### REAGENTS AND MATERIALS

Reagent-grade cyclohexane
Naphthalene
2M sodium chloride solution
Ice
Assigned unknown solids for molar mass determination

## BACKGROUND

### Associated Readings

The following sections of *Chemical Principles* should be read in conjunction with Experiment 11.

**Chapter 12** The Properties of Solutions
12.5 Boiling Point and Freezing Point of Solutions (The Freezing Behavior of Solutions) See also Example 12.12

In a dilute solution, the vapor pressure of the solvent A over the solution ($P_A$) is approximately equal to the mole fraction of A present in solution ($X_A$) times the vapor pressure of pure A ($P_A°$). This is known as **Raoult's Law**.

$$P_A = X_A \cdot P_A°$$

The law involves only the mole fraction of solvent (and thus of solute), but is independent of the chemical identity of the solute, though the usual distinctions between electrolytes and nonelectrolytes must be observed in calculating mole fractions. Properties of solutions which depend on the amount (in moles) of solute but not on the chemical identity of the solute are known as colligative properties. Vapor pressure is a **colligative property**.

Raoult's law applies strictly only to ideal solutions, but is approximated reasonably well for most organic solvents in dilute solutions of nonelectrolytes.

The lowering of vapor pressure of a pure liquid following addition of a nonvolatile solute means that the boiling and freezing points of the solution are not the same as those of the pure liquid. The boiling point is raised by an amount $\Delta T_b$, and the freezing point is lowered by an amount $\Delta T_f$. Both $\Delta T_b$ and $\Delta T_f$ are colligative properties, being dependent on how much the vapor pressure is lowered, which in turn depends only on the amount of added solute but not on its nature.

The way in which vapor pressure changes as a function of temperature is different for different solvents, so the magnitudes of $\Delta T_b$ and $\Delta T_f$ depend on the particular solvent used as well as on its concentration. As an approximation, for dilute solutions we write

$$\Delta T_f = K_f \cdot m,$$

where $m$ is the molal concentration of solute and $K_f$ is the molal freezing point depression constant. If $\Delta T_f$ is given in K and $m$ is expressed as moles of *solute* per kilogram of solvent, K has dimensions K-kg/mol. An analogous relationship holds for $\Delta T_b$, with $K_b$ being the molal boiling point elevation constant.

Because the vapor pressure of most liquids rises sharply as the boiling point is approached, observed values of $K_b$ are small, few exceeding 5 K-kg/mol. Values of $K_f$, on the other hand, may be quite large, e.g., 31.8 K-kg/mol for carbon tetrachloride. The measured freezing point depressions for dilute solutions are thus a convenient way of establishing the molal concentration of an unknown solute, and hence its molar mass.

In this experiment the value of $K_f$ for cyclohexane is determined by measuring the freezing point depression produced by a known concentration of solute of known molar mass. The value of $K_f$ found is used to find the molal concentration of a known mass of solute of unknown molar mass by a similar measurement of $\Delta T_f$, and hence the molar mass is found.

### Cooling Curves

To determine the freezing point of a liquid correctly, we need to consider the variation of temperature with time as the liquid is steadily cooled. This is illustrated by Figures 11.1 and 11.2.

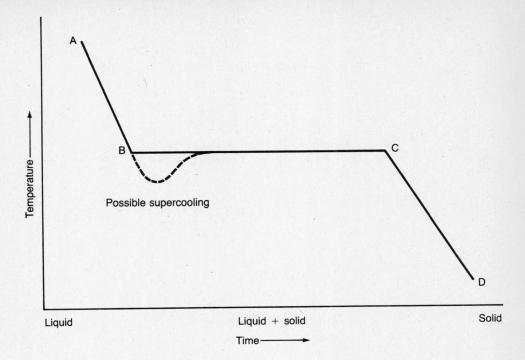

**Figure 11.1**
Cooling curve for a pure liquid.

Figure 11.1 shows the behavior of a pure liquid. The line AB shows the drop in temperature as the liquid is steadily cooled. At point B, crystals of the solid phase appear and in the range BC solid and liquid coexist, with the amount of liquid steadily decreasing. In the range BC latent heat is released to the surroundings but the temperature of the solid-liquid mixture remains constant. This constant temperature is the freezing point of the liquid. At point C all of the liquid has turned to solid and in the range CD the solid steadily cools. In some cases, as indicated by the dotted line, supercooling of the liquid may occur before solid forms.

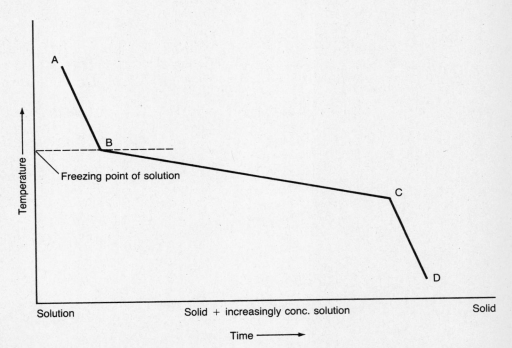

**Figure 11.2**
Cooling curve for a solution.

For a solution the situation is more complex. Again in the range AB the liquid cools steadily, but in the range BC where both solid and liquid coexist, the temperature initially falls and may continue to do so throughout the range. This happens because as liquid solvent is converted to solid, the molal concentration of the solution increases and the freezing point of the solution drops to reflect the increase in $m$. Eventually the solid may become saturated, in which case no further drop in temperature takes place until point C where all of the solvent is now solidified. The solvent and solute may crystallize separately or the solution may freeze as a solid of definite composition, a so-called **eutectic solid.**

The freezing point temperature needed for this experiment is the temperature at point B, found graphically as the intersection of lines AB and CB. Note that in Figure 11.2 the difference in the slopes of AB and CB has been exaggerated somewhat to emphasize the points made above.

## PROCEDURE

### SAFETY NOTES
Cyclohexane is a volatile flammable solvent. Keep open flames away from its vapors. Dispose of waste cyclohexane only in the labelled containers provided.

Take care handling the narrow-range thermometers. Check that the mercury column is unbroken and that the thermometer registers close to 0°C when immersed in the ice-water mixture. If the mercury thread is broken, have the thermometer replaced at the stockroom. Do not try to shake the thermometer as you would a clinical thermometer. If a thermometer is broken, take care to gather up all the spilled mercury and take it to the stockroom for proper disposal. Never discard spilled mercury in the trash containers or down the sink. Always inform your instructor of any mercury spill.

### 1. Preparing the Cooling Mixture

Fill a 1-L beaker or other suitable container to the brim with crushed ice. Fill the beaker to within a few centimeters of the rim with the $2M$ NaCl solution, taking care not to overfill. With a clamp and stand support the outer tube of the freezing point apparatus so that as much as possible of the tube is immersed in the coolant. Carefully clean up any brine spill that arises now or later.

### 2. Determining the Freezing Point of Cyclohexane

Cyclohexane, $C_6H_{12}$, freezes around 6.5°C and has a $K_f$ of around 20 K-kg/mol. The exact values of these two quantities are determined in this experiment. To obtain good results, the inner glass tube, the stirring wire, and the thermometer must be completely clean else the cyclohexane will be contaminated. Before beginning the experiment, wash out the tube with a little hot water and detergent, and do the same for the stirrer. Rinse both thoroughly with distilled water and invert the tube to drain on a dry towel. Rinse the tube with two 5-mL portions of acetone and again let it drain completely dry. During this stage of the procedure handle the tube as little as possible and then with a dry towel. Clean and dry the thermometer. Repeat the cleaning process before each use of the tube and again at the end of the experiment.

Place the inner tube in a clean, dry 100-mL beaker and find the mass of the assembly to the nearest milligram, using the fine balance. Add about 25 mL of cyclohexane to the tube and hold the tube in the freezing mixture until crystals of the solid form. Remove the tube from the cooling mixture, dry the outside completely, and let it warm up slightly until the crystals initially formed have melted. Place the tube in the outer jacket of the apparatus, insert the thermometer and stirrer, and determine the freezing point of cyclohexane by constructing a cooling curve like that in Figure 11.1. Make temperature readings every 30 seconds for a period between 5 and 10 minutes, but in any event continue until all the cyclohexane has solidified. Stir the liquid once each second with a gentle up-and-down motion of the stirrer. Read the thermometer as accurately as possible and enter your readings directly on the graph paper. At the end of this step, remove the inner tube and allow the solid to melt.

### 3. Determination of $K_f$ for Cyclohexane

On the rough balance weigh out about 0.35 g of naphthalene in a clean, dry 30-mL beaker. Carefully dry off any condensation on the outside of the inner tube, place it in the same clean, dry 100-mL beaker as before and use the fine balance to find the mass of cyclohexane in the tube. Quickly transfer the naphthalene to the inner tube containing the cyclohexane and reweigh to find the exact mass of solid added. Make these weighings as quickly as possible, as cyclohexane is quite volatile. Replace the thermometer and stirrer in the inner tube and stir to dissolve all of the naphthalene and obtain a homogeneous solution.

Place the inner tube directly in the freezing mixture until crystallization begins, remove and dry it, allow the solid formed to redissolve, and place the inner tube inside the outer as before. Plot a cooling curve for the solution and determine the freezing point.

### 4. Determination of the Freezing Point Depression Produced by an Unknown Solid

Discard the cyclohexane-naphthalene solution in the marked disposal containers. Thoroughly clean and dry the inner tube, thermometer and stirrer. Using about 0.35 g of the assigned unknown and about 20 mL of cyclohexane, repeat the procedure described in the last section to find the concentration and freezing point of the unknown.

Properly dispose of leftover solution and thoroughly clean the apparatus before returning it to the stockroom.

Name _____  Instructor _____

## PRELABORATORY PREPARATION SHEET

1. What is meant by the term "colligative property"?

   _____

   _____

2. Naphthalene has the molecular formula $C_{10}H_8$. Cyclohexane has a density of 0.779 g/cm$^3$. Calculate the molal concentration of a solution of 0.35 g naphthalene in 25 mL cyclohexane.

   _____

   _____

   _____

3. Calculate the magnitude of the freezing point depression expected for the above solution, assuming $K_f$ for cyclohexane to be about 20 K-kg/mol.

   _____

   _____

   _____

4. Assuming the melting point of cyclohexane to be around 6.5°C, prepare:
   (a) A sheet of graph paper, labeled for temperature and time, suitable for recording a cooling curve for pure cyclohexane. Assume temperature readings accurate to 0.05°C and readings every 30 seconds.
   (b) A similar labeled graph for a cooling curve for the solution of naphthalene in cyclohexane.
   (c) A similar labeled graph for the solution of the unknown in cyclohexane. Assume the unknown to have a molar mass of 150.

   Present these graphs for inspection by your instructor at the start of the laboratory period. Remember that you will need a sharp No. 3H, or harder, pencil to enter points on the graphs.

Name _____  Instructor _____

Partner _____

## DATA SHEET

Students work in pairs, one reading time, the other temperature. The measured temperatures are to be entered directly onto the labeled graphs, not jotted down on scratch paper for later copying. In this way you can visually follow the cooling process as it takes place. One member of the team may enter the data points on both graphs, but you work independently in interpreting the data and in preparing the report sheet.

**I**

Mass of beaker + tube + cyclohexane  _____

Mass of beaker + tube  _____

Mass of cyclohexane  _____

Mass of beaker + tube + solvent + $C_{10}H_8$  _____

Mass of beaker + tube + solvent  _____

Mass of $C_{10}H_8$  _____

**II**

Mass of beaker + tube + cyclohexane  _____

Mass of beaker + tube  _____

Mass of cyclohexane  _____

Mass of beaker + tube + solvent + unknown  _____

Mass of beaker + tube + solvent  _____

Mass of unknown  _____

235

Name _____  Instructor _____

Partner _____

## DATA SHEET

Students work in pairs, one reading time, the other temperature. The measured temperatures are to be entered directly onto the labeled graphs, not jotted down on scratch paper for later copying. In this way you can visually follow the cooling process as it takes place. One member of the team may enter the data points on both graphs, but you work independently in interpreting the data and in preparing the report sheet.

**I**

Mass of beaker + tube + cyclohexane _____

Mass of beaker + tube _____

Mass of cyclohexane _____

Mass of beaker + tube + solvent + $C_{10}H_8$ _____

Mass of beaker + tube + solvent _____

Mass of $C_{10}H_8$ _____

**II**

Mass of beaker + tube + cyclohexane _____

Mass of beaker + tube _____

Mass of cyclohexane _____

Mass of beaker + tube + solvent + unknown _____

Mass of beaker + tube + solvent _____

Mass of unknown _____

Name _____  Instructor _____

Partner _____

## REPORT SHEET

This report is to be completed by each student working independently. Attach your cooling curve graphs to the report sheet.

1. Give the freezing point of cyclohexane derived from your cooling curve. _____
2. Calculate the molal concentration of your naphthalene solution.

   _____

   _____

3. Give the freezing point of the naphthalene solution derived from your cooling curve. _____

4. Enter the freezing point depression. _____
5. Calculate $K_f$ for cyclohexane.

   _____

   _____

   _____

6. Give the freezing point of the unknown solution. _____

7. Enter the freezing point depression for that solution. _____
8. Calculate the molal concentration of the unknown.

   _____

   _____

   _____

9. Give the mass of unknown used. _____
10. Calculate the molar mass of the unknown.

    _____

239

## WORKSHEET

Carry out all the calculations for Experiment 11 on this sheet, attach it to your report sheet, and submit the two together. Do your calculations directly on this sheet. Do not do them on scratch paper and then copy them on to this sheet. Make no erasures and delete incorrect entries with a single line so as to leave the original data legible. Use a pen.

# EXPERIMENT 12

# The Solubility Product of Calcium Iodate

### AIM

To find the solubility product of the sparingly soluble salt calcium iodate, $Ca(IO_3)_2$, by determining the equilibrium iodate concentration in a saturated solution of the salt by redox titrimetry.

### EQUIPMENT

From stockroom:   5-mL and 25-mL pipets
                        50-mL buret
                        250-mL volumetric flask
From locker:      Three Erlenmeyer flasks (200-, 250- or 300-mL)
                        30-, 50-, and 100-mL beakers
                        Glass filter funnel
                        500-mL glass bottle
                        Washbottle

## REAGENTS

Sodium thiosulfate stock solution, $Na_2S_2O_3$, 0.025$M$
Potassium dichromate, $K_2Cr_2O_7$
Potassium iodide, KI
1$M$ Sulfuric acid
Starch indicator
Calcium iodate, saturated solution

## BACKGROUND

### Associated Readings

The following sections of *Chemical Principles* should be read in conjunction with Experiment 12.

**Chapter 14** Chemical Equilibrium
        14.1 The Equilibrium State and the Equilibrium Constant
        14.5 Solutions of Sparingly Soluble Substances: The Solubility Product
            See Table 14.3 Solubility Products of Some Sparingly Soluble Ionic Salts at Room Temperature
**Chapter 16** Oxidation-Reduction
        16.1 Oxidation-Reduction Reactions
        16.2 Oxidation-Reduction Processes in Aqueous Solution (Oxidation-Reduction Titrations)

---

The solubility constant of a sparingly soluble salt is the equilibrium constant associated with the equilibrium between the solid salt and its constituent ions in solution. For calcium iodate, $Ca(IO_3)_2$, this equilibrium is represented by the equation

$$Ca(IO_3)_2(s) \rightleftharpoons Ca^{2+}(aq) + 2IO_3^-(aq)$$

and the equilibrium constant, $K_{sp}$, is given by

$$K_{sp} = [Ca^{2+}][IO_3^-]^2$$

The concentration of the solid phase, being a constant, does not appear in the expression for $K_{sp}$.

$K_{sp}$ is readily evaluated for a solution of calcium iodate if the equilibrium concentration of either ionic species is known and no other salt containing either ion is present in the same solution. In this experiment, which is based on a series of redox titrations, the equilibrium concentration of the iodate will be determined.

This will be done by using the iodate ion in a known volume of saturated calcium iodate solution to oxidize iodide ion to free iodine in acid solution according to the equation

$$IO_3^-(aq) + 5I^-(aq) + 6H^+(aq) \longrightarrow 3I_2(aq) + 3H_2O \tag{1}$$

The amount of free iodine generated in this way is determined by titration against standard thiosulfate solution:

$$I_2(aq) + 2S_2O_3^{2-}(aq) \longrightarrow 2I^-(aq) + S_4O_6^{2-}(aq) \tag{2}$$

Because sodium thiosulfate—the usual source of thiosulfate ion for volumetric work—normally forms a hydrate of indeterminate water content, it is not suitable for use as a primary standard. The concentration of the stock thiosulfate solution must therefore be established by titration. A solution of potassium dichromate, a primary standard, is prepared in known concentration and is used to oxidize a known amount of iodide ion to iodine according to the equation

$$Cr_2O_7^{2-}(aq) + 6I^-(aq) + 14H^+(aq) \longrightarrow 2Cr^{3+}(aq) + 3I_2(aq) + 7H_2O \qquad (3)$$

The concentration of thiosulfate is then found by measurement of the amount of solution needed to reduce the iodine thus formed back to iodide, following reaction (2).

In carrying out the iodine determinations, the iodine solutions must be stabilized to minimize losses caused by air oxidation and evaporation of the iodine. This is done by adding a large excess of iodide ion to take advantage of the equilibrium

$$I_2(aq) + I^-(aq) \rightleftharpoons I_3^-(aq)$$

for which the equilibrium constant

$$K = [I_3^-]/[I_2][I^-]$$

is about 550 at room temperature. By ensuring that a very high concentration of iodide is always present, the amount of free iodine in solution is kept to a minimum. As the titration proceeds, iodine is consumed and the equilibrium concentrations of the various species shift, in accordance with Le Chatelier's principle, until all the tri-iodide ion initially formed is reconverted to iodine and iodide.

## PROCEDURE

### SAFETY NOTES

Dichromate ion contains chromium in oxidation state +6. In that state chromium is a suspected carcinogen. No health hazard is posed by the operations involving dichromate in this experiment, but all dichromate residues and solutions must be properly disposed of. All leftover dichromate solution must be placed in the specially marked container, and all spills of solid dichromate must be carefully cleaned up and the solid placed in the same container. These materials will later be reduced to Cr(III), a harmless species.

### 1. Preparation of Standard Dichromate Solution

Dichromate solutions place great demands on the cleanliness of your glassware. Unless volumetric flasks, pipets, and burets are scrupulously cleaned, dichromate will bead on their surfaces. Accurate results are impossible to achieve if that occurs, so it pays to be more than usually assiduous in cleaning the volumetric ware. Glassware should be cleaned with distilled water and detergent solution, then rinsed with distilled water. Make sure all items drain cleanly and no drops stick to the inside surfaces.

On the rough balance, weigh out about 1 g of solid potassium dichromate into a clean, dry 30-mL beaker. Transfer the beaker and its contents to the fine balance and weigh to the nearest milligram. Place a clean, dry glass filter funnel in the mouth of the 250-mL volumetric flask and transfer the dichromate from the beaker to the flask through the funnel. Do not worry about getting the last drop of dichromate out of the beaker but instead weigh the empty beaker, again to the nearest milligram, to determine the exact mass of solid transferred. Use distilled water from your washbottle to wash out any dichromate adhering to the funnel into the flask. Remove the funnel and, again, with the washbottle wash any droplets adhering to the end of the funnel into the flask. Be liberal with these rinses so that all the dichromate ends up in the flask.

With distilled water make up the volume of the flask to about 150 mL and swirl to dissolve all the dichromate. Fill the flask with distilled water to about 1 cm below the mark. Make sure no droplets adhere to the inside of the neck and, with the washbottle, make up the solution to the mark by dropwise addition of distilled water. Stopper the flask and invert it repeatedly to ensure homogeneity of the solution.

## 2. Standardization of the Thiosulfate Solution

Prepare a solution of potassium iodide by adding 45 mL of distilled water to 18 g of the iodide contained in a clean 100-mL beaker. Cover the beaker with a watchglass to minimize oxidation by the air.

Collect about 60 mL of $1M$ sulfuric acid in another clean 100-mL beaker and about 20 mL of starch indicator solution in a clean 50-mL beaker.

Clean and thoroughly rinse three Erlenmeyer flasks (200-, 250- or 300-mL) and add about 15 mL of the $1M$ sulfuric acid to each.

Thoroughly clean the 25-mL pipet with cold distilled water and detergent. Rinse the pipet with distilled water. Transfer about 30 mL of the dichromate solution to a clean, dry 100-mL beaker and use it to make three rinses of the pipet. Make sure that no droplets of dichromate adhere to the inner surfaces of the pipet after the rinsing. If droplets are present, reclean the pipet properly before going on, repeating the dichromate rinses.

Discard any residual dichromate from the beaker into the specially marked disposal container. Refill the beaker from the volumetric flask and pipet exactly 25 mL of dichromate into each Erlenmeyer flask. Add about 5 mL of iodide solution to each flask and swirl the contents gently to promote mixing. Invert a small beaker over the neck of each flask so as to minimize air oxidation. The flasks should stand for about 10 minutes to allow the formation of tri-iodide ion to go to completion, indicated by the development of a brown color in the solution.

While that reaction is proceeding, ready a buret with the thiosulfate solution in the usual way. Record the initial reading on the buret.

Now titrate the iodine solution against the thiosulfate. Use the color of the solution as a guide to the approaching end point. Add the thiosulfate, about 3 mL at a time, with gentle swirling to ensure thorough mixing, up to about 20 mL and then dropwise until the brown color begins to fade. At this point, add 5 mL of the starch solution and observe the formation of the intense blue color of the starch-iodine complex. With your washbottle rinse down the inside surfaces of the flask to ensure that all the reagents are in the body of the solution and continue with a dropwise addition of thiosulfate until the deep blue color is discharged and the solution takes on the characteristic light blue color of the hydrated $Cr^{3+}$ ion. Record the final buret reading.

Quickly repeat the titration on the remaining two samples. In all cases read the buret to the nearest 0.02 mL and average only the results of titrations agreeing to within 0.10 mL.

Carry out a blank titration by cleaning one of the flasks and adding to it 25 mL of distilled water in place of the dichromate solution. Add 15 mL of the $1M$ sulfuric acid to the flask, followed by 5 mL of iodide solution and 5 mL of starch indicator. Cover the flask neck with a small beaker and let stand for 10 minutes to mimic the previous system. Development of an intense blue color indicates the presence of iodine from air oxidation. In this case do a drop-by-drop titration with thiosulfate to discharge the blue color and subtract the volume of thiosulfate taken from the mean volume needed to react with the dichromate solutions in the standardization. If no blue color develops in the blank, proceed directly to Step 3.

## 3. Determination of Iodate Concentration

Half fill a clean, dry 50-mL beaker with the stock solution of saturated calcium iodate. Transfer part of this solution to a second clean, dry beaker and use it to rinse the previously cleaned 5-mL pipet.

Collect about 60 mL of 1$M$ sulfuric acid and 20 mL of starch indicator.

Thoroughly clean the three flasks used in the previous step, rinsing them with distilled water. Into each flask pipet 5 mL of the iodate solution, add 30 mL of distilled water and 15 mL of the sulfuric acid. Swirl gently to mix the contents while adding 5 mL of the iodide solution to each flask. Cover the neck of each flask with an inverted beaker and let stand for about 10 minutes to allow the reaction to go to completion.

Refill the buret with thiosulfate solution and titrate each sample to a colorless endpoint by following the same procedure as in Step 2. A much smaller volume of thiosulfate will be needed in these titrations.

Once again carry out a blank titration substituting 5 mL of distilled water for the iodate solution, and again subtract the volume of thiosulfate required in the blank from the mean volume needed for the three titrations.

## 4. Final Things

Clean and rinse the volumetric glassware as soon as you have completed Step 3, and return it to the stockroom. Be sure that all dichromate leftovers are disposed of properly. On no account flush dichromate solutions or solids into the drain system.

Name _____  Instructor _____

## PRELABORATORY PREPARATION SHEET

1. In pipetting a standard solution, why do you suppose you should not pipet directly from the volumetric flask?

2. What is the purpose of the blank titration in Steps 2 & 3?

3. Why do you not need an exact measure of the amount of iodide added in Steps 2 & 3?

4. Why is cold distilled water used in the cleaning and rinsing of volumetric glassware?

5. Complete the following stoichiometric equivalences by providing the missing coefficient on the right-hand side corresponding to one mole of the species on the left.

   $Cr_2O_7^{2-} \equiv$ _____ $I_2$    $I_2 \equiv$ _____ $S_2O_3^{2-}$

   $IO_3^- \equiv$ _____ $I_2$    $Cr_2O_7^{2-} \equiv$ _____ $S_2O_3^{2-}$

   $IO_3^- \equiv$ _____ $S_2O_3^{2-}$

6. If you have not previously carried out my volumetric work, e.g. titrations, read the Appendix for more details of the various operations required.

Name _____   Instructor _____

## DATA SHEET

1. Dichromate standardization:

    Mass of beaker + dichromate  _____

    Mass of beaker alone  _____

    Mass of dichromate delivered  _____

2. Titrations (25 mL dichromate taken):

    Final buret reading  _____  _____  _____

    Initial buret reading  _____  _____  _____

    Volume of thiosulfate delivered  _____  _____  _____

    Mean volume of thiosulfate  _____ ± _____

    Blank titration  _____

    Corrected mean volume  _____ ± _____

3. Iodate titrations (5 mL iodate taken):

    Final buret reading  _____  _____  _____

    Initial buret reading  _____  _____  _____

    Volume of thiosulfate delivered  _____  _____  _____

    Mean volume of thiosulfate  _____ ± _____

    Blank titration  _____

    Corrected mean volume  _____ ± _____

Name _____  Instructor _____

# DATA SHEET

1. Dichromate standardization:

    Mass of beaker + dichromate  _____

    Mass of beaker alone  _____

    Mass of dichromate delivered  _____

2. Titrations (25 mL dichromate taken):

    Final buret reading  _____  _____  _____

    Initial buret reading  _____  _____  _____

    Volume of thiosulfate delivered  _____  _____  _____

    Mean volume of thiosulfate  _____ ± _____

    Blank titration  _____

    Corrected mean volume  _____ ± _____

3. Iodate titrations (5 mL iodate taken):

    Final buret reading  _____  _____  _____

    Initial buret reading  _____  _____  _____

    Volume of thiosulfate delivered  _____  _____  _____

    Mean volume of thiosulfate  _____ ± _____

    Blank titration  _____

    Corrected mean volume  _____ ± _____

Name _____  Instructor _____

## REPORT SHEET

1. Give your calculation of the molarity of your dichromate solution.

   _____

   _____

   _____

   _____

2. Calculate the moles of dichromate in 25 mL of your solution.

   _____

3. Give the moles of iodine to which this corresponds.

   _____

4. Give the moles of thiosulfate to which this corresponds.

   _____

5. From the mean volume of thiosulfate required in the dichromate titration, and using the answers to 2, 3, & 4 above, calculate the molarity of the thiosulfate solution.

   _____

   _____

   _____

6. From the mean volume of thiosulfate required for the iodate titration, calculate the moles of thiosulfate required for 5 mL of iodate solution.

   _____

7. Give the moles of iodate to which this thiosulfate corresponds.

   _____

8. Calculate the molarity of the iodate solution.

   _____

9. What is the molarity of calcium in the calcium iodate solution?

   _____

10. From your answers to 8 & 9 above, evaluate $K_{sp}$ for calcium iodate.

    _____

11. Use the uncertainties in the various experimental steps to estimate the likely error in your calculation of $K_{sp}$.

    _____

    _____

    _____

255

## WORKSHEET

Carry out all the calculations for Experiment 12 on this sheet, attach it to your report sheet, and submit the two together. Do your calculations directly on this sheet. Do not do them on scratch paper and then copy them on to the sheet. Make no erasures and delete incorrect entries with a single line so as to leave the original data legible. Use a pen.

# EXPERIMENT 13

# Determination of an Equilibrium Constant By Spectrophotometry

## AIM

To find the room temperature equilibrium constant for the formation of a metal ion complex. To determine by absorption spectrophotometry the concentration of the complex when equilibrium is reached from a range of starting concentrations of reactants.

## EQUIPMENT

From locker:  Two 50-mL beakers
Two 100-mL beakers
Test tubes
25-mL graduated cylinder
Laboratory: 5-, 10-, and 50-mL graduated cylinders
Spectrophotometer and sample cell

## REAGENTS

$0.20 M$ stock solution of $Fe(NO_3)_3$
$4.0 \times 10^{-4} M$ stock solution of KSCN
Distilled water

## BACKGROUND

### Associated Readings

The following sections of *Chemical Principles* should be read in conjunction with Experiment 13.

**Chapter 8**  Molecular Structure and Stability
   8.1 Experimental Methods (Spectroscopy)
**Chapter 14** Chemical Equilibrium
   14.1 The Equilibrium State and the Equilibrium Constant
   14.2 The Principle of Le Chatelier
**Chapter 18** Coordination Chemistry
   18.5 Complex Ions in Aqueous Solution

---

Chemical reactions always proceed in two directions simultaneously, but initially at different rates. As soon as products are formed from reactants, the process of reconversion to reactants begins.

The laws governing the rates of chemical reactions show dependance on the concentration of reactants and are often complex, depending on the particular reaction mechanism involved. The rate of any reaction is not predictable but can only be established experimentally. However, in all chemical processes there comes a time at which the rates of the forward and reverse reactions are the same. When this point is reached, the system is said to be at **equilibrium.** In some systems equilibrium is attained very rapidly indeed, in others it is reached extremely slowly. Just as rates of reaction are dependent on temperature, so is the exact position of equilibrium different at different temperatures.

Once the system has reached equilibrium, however, at a given temperature, an experimental law defines the relationship between the concentrations of reactants and products. This law, the **Law of Mass Action,** may be illustrated by the generalized chemical reaction whose balanced equation is

$$a\text{A} + b\text{B} + \ldots \rightleftharpoons x\text{X} + y\text{Y} + \ldots$$

The law states that at equilibrium a constant, $K$, exists equal to

$$\frac{[\text{X}]^x[\text{Y}]^y \ldots}{[\text{A}]^a[\text{B}]^b \ldots}$$

which defines the relationship between the equilibrium concentrations of reactants and products. For reactions taking place in solution, the square brackets indicate molar concentrations and each is raised to a power equal to that indicated by the stoichiometric coefficient for the species in the balanced equation. Species whose concentrations do not change during the reaction, such as those of pure solids or liquids, are not included in the expression for $K$, the equilibrium constant.

As an example, $K$ for the equilibrium involving the oxidation of iron(II) to iron(III) by permanganate ion in acidic aqueous solution, according to the equation

$$5\text{Fe}^{2+}(aq) + \text{MnO}_4^-(aq) + 8\text{H}^+(aq) \rightleftharpoons 5\text{Fe}^{3+}(aq) + \text{Mn}^{2+}(aq) + 4\text{H}_2\text{O}$$

is

$$K = \frac{[\text{Fe}^{3+}]^5[\text{Mn}^{2+}]}{[\text{Fe}^{2+}]^5[\text{MnO}_4^-][\text{H}^+]^8}$$

Note that the value of $K$ at any given temperature does not depend on the initial concentrations of reactants and products. It tells us only the relationship among the concentrations once equilibrium has been attained. However, for a particular set of initial concentrations of reactants and products, the value of $K$ for a given reaction allows us to predict in which direction the reaction will proceed to reach equilibrium, and it tells us just how far the reaction will proceed under these circumstances. It tells us nothing about how fast either event will occur.

In a qualitative sense, the magnitude of $K$ has immediate predictive value. Reactions having very large values of $K$ reach equilibrium with very low concentrations of reactants relative to products. Such reactions are often said to go to completion. Though some small amount of reactant will inevitably remain at equilibrium, it is often too small to be detectable. Conversely, reactions with very small values of $K$ form very low concentrations of products and are often said not to proceed to any appreciable extent. An example is the autoionization of water where the molar concentration of hydrated protons at equilibrium is only $1 \times 10^{-7} M$.

Because $K$ is independent of the initial concentrations of reactants and products, once it has been found for an equilibrium arising from a specific set of initial conditions, it is also applicable to the equilibria arising from any other set of initial concentrations at the same temperature. The actual concentrations of reactants and products at equilibrium may vary from experiment to experiment but, for the same reaction at the same temperature, always adjust themselves to conform to the relationship indicated by the equilibrium constant expression and yield always the same numerical value for $K$.

In this experiment we make numerical determinations of the equilibrium constant for a reaction that reaches equilibrium very quickly. We do so for a range of initial reactant concentrations and confirm the assertions made in the previous paragraph.

The reaction we will study is the replacement of a water molecule attached to a $Fe^{3+}$ ion, in aqueous solution, by a thiocyanate ion according to the equation

$$Fe(H_2O)_6^{3+}(aq) + SCN^-(aq) \rightleftharpoons [Fe(H_2O)_5SCN]^{2+}(aq) + H_2O$$

The equilibrium constant for this reaction is

$$K = \frac{[Fe(H_2O)_5SCN^{2+}]}{[Fe(H_2O)_6^{3+}][SCN^-]}$$

Further successive substitutions of $SCN^-$ for $H_2O$ do occur, such as

$$[Fe(H_2O)_5SCN]^{2+}(aq) + SCN^-(aq) \rightleftharpoons [Fe(H_2O)_4(SCN)_2]^+(aq) + H_2O$$

but $K$ for this reaction is very much smaller than that for the initial substitution and by using initial concentrations of $Fe^{3+}$ very much greater than those of $SCN^-$ we can minimize their effects. (Recall Le Chatelier's Principle.)

Let us simplify writing the equilibrium constant by omitting the water molecules so that we have

$$K = [FeSCN^{2+}]/[Fe^{3+}][SCN^-]$$

To evaluate $K$ we need to know all three molar concentrations at equilibrium. Three separate determinations are unnecessary, however, as the three are linked. For instance, if $C_M$ is the initial molar concentration of the metal ion then its concentration at equilibrium is given by

$$[Fe^{3+}] = C_M - [FeSCN^{2+}]$$

Similarly, if $C_L$ is the initial ligand concentration, then at equilibrium

$$[SCN^-] = C_L - [FeSCN^{2+}]$$

since, by the stoichiometry of the reaction, one complex ion is formed by combination of one $Fe^{3+}$ ion and one $SCN^-$ ion. With $C_M$ and $C_L$ known, only $[FeSCN^{2+}]$ remains to be found.

The complex ion is highly colored whereas the visible absorbances of $Fe^{3+}$ and $SCN^-$ are negligible in comparison. The absorbance of a known concentration of the

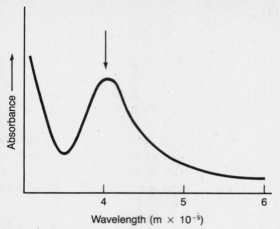

**Figure 13.1**
A typical absorption spectrum showing a maximum at 415 nm.

complex ion is found by spectrophotometry and we apply the law linking absorbance and concentration to establish the equilibrium concentration of the ion when equilibrium is reached from a range of starting concentrations of $Fe^{3+}$ and $SCN^-$.

## Spectrophotometry

Colored substances absorb radiation in the visible region of the electromagnetic spectrum. A red solution of $FeSCN^{2+}$ absorbs toward the blue end of the visible region, the remaining red wavelengths being transmitted. Absorption varies as a function of wavelength and this variation is measured by an instrument known as an absorption spectrophotometer. A typical spectrum is shown in Figure 13.1. Absorption spectra are often characteristic of a particular substance or of one of its components or constituents.

At a particular wavelength the light absorbed, the so-called **absorbance,** $A$, is defined in terms of $I_o$, the intensity of light incident on the sample, and $I$, the intensity of light transmitted, according to the equation

$$A = \log(I_o/I)$$

When 100% of the incident light is transmitted ($I = I_o$, $A = \log 1$) there is zero absorbance. When only half of the incident light is transmitted, $I_o/I = 2$ and $A = 0.3$. With 10% **transmission** (90% **absorption**) $I_o/I = 10$ and $A = 1$. With only 1% transmission (99% absorption), the practical limit for most instruments, $A = 2$. If an absorbance greater than 2 is found, the solution should be diluted to bring the absorbance within an accurately measurable range.

Absorbance is proportional to concentration.

$$A = \epsilon bc$$

where $\epsilon$ is the **molar absorptivity,** $b$ is the path length of light through the system, measured in cm, and $c$ is the molar concentration of the substance under study. $A$ is dimensionless so that $\epsilon$ has units of $mol^{-1}\ cm^{-1}$.

In a given spectrophotometer, the path length $b$ is fixed so that, at any given wavelength, the product $\epsilon b$ may be determined from the measured absorbance by a solution of known concentration. With $\epsilon b$ known, any unknown concentration is found by measuring the absorbance of the solution at the same wavelength.

We determine $\epsilon b$ by measuring $A$ for the solution that results from the first mixing using the undiluted stock $Fe^{3+}$ solution. The large excess of $Fe^{3+}$ drives the reaction effectively to completion giving a $0.20mM$ solution of the complex ion.

# Determination of an Equilibrium Constant by Spectrophotometry

## PROCEDURE

### SAFETY NOTES
There are no obvious hazards in this experiment. Dispose of the leftover reagent solutions by flushing them down the sink with cold water.

### 1. Calibrating the Spectrophotometer

There are several different models of spectrophotometer in use. The directions given here apply to the Spectronics Model 20 instrument, widely used in introductory laboratories. If your laboratory has a different model, your instructor will advise you on its use.

The spectrophotometer will normally be turned on and ready to use. Obtain a sample cell matching the standard cell provided with the instrument and fill both cells two-thirds full of distilled water. Close the sample holder cover and set the wavelength to 450 nm. The needle should indicate zero transmittance or infinite absorbance—be sure to distinguish between the two scales. If it does not, adjust the left hand knob on the front of the instrument until an infinite reading is obtained for $A$. Now raise the sample cover and insert the *standard* cell containing water. A zero absorbance reading should result. If not, adjust to zero with the right hand knob. Only a small adjustment, if any, should be needed. In case of difficulty consult your instructor.

Now remove the standard cell and replace it with your sample cell containing water. Mark the sample cell on its top edge so that it can always be inserted into the instrument in the same orientation for each subsequent reading. Ideally, the absorbance reading should be zero, but if not, note the small difference from zero and subtract this amount from each later reading to ensure a consistent baseline matching that of the instrument itself.

### 2. Preparation of Reagent Solutions

The stock solution of ferric iron is $0.20M$ ($200mM$) $Fe(NO_3)_3$ in water. It is used at this concentration and diluted by factors of 10, 20, and 40.

Take about 20 mL of the stock solution in a clean, dry labeled test tube. Prepare a $20mM$ solution by taking 5 mL of this stock sample and diluting it to 50 mL using graduated cylinders for the measurements. Measure out 25 mL of the $20mM$ solution in graduated cylinder and transfer the remainder to a clean, dry labeled 50-mL beaker. Dilute the 25 mL in the cylinder to 50 mL to prepare a $10mM$ solution. Measure 25 mL of the $10mM$ solution into a graduated cylinder and transfer the remainder to a clean, dry labeled beaker, as before. Dilute the $10mM$ solution in the same way to prepare a $5mM$ solution, storing it in another clean labeled beaker.

The thiocyanate solution is $0.40mM$ KSCN. It is used at this concentration and diluted by a factor of two. Take about 60 mL of the stock solution in a clean, dry 100-mL beaker.

### 3. Carrying out the Reaction and Measuring Absorbance

For each determination, 10 mL of one of the various $Fe^{3+}$ solutions is mixed with 10 mL of one or other of the thiocyanate solutions to yield a final reaction volume of 20 mL. The data sheet shows the combinations required. Where the $0.20mM$ thiocyanate is needed, use 5 mL of the stock solution and 5 mL of distilled water. Use graduated cylinders for the measurements and a test tube for the mixing. Be sure to thoroughly mix the reagents in each case by placing your thumb on the top of the test tube and inverting the tube several times. Be sure that the test tube used for the mixing is as clean and dry as possible.

To measure the absorbance, wash your experimental spectrophotometer cell with distilled water, rinse it once with a little of the solution to be measured and then fill it two-thirds full of the solution and insert it in the proper orientation into the sample holder. Record the absorbance at 450 nm on the data sheet.

The first mixing, using the undiluted stock $Fe^{3+}$ solution, gives A for a 0.20m$M$ solution of the complex ion, it being assumed that the large excess of iron drives the reaction well to the right. This value may be used to establish the value of the product $\epsilon b$, so that the concentration of complex can be derived for the other mixings. You should carry out two additional mixings of the two reagents at concentrations different from those used in the first six. The only restriction is that the absorbance at equilibrium should lie in the useful range of the machine.

## 4. Analysis of Results

The experiment leads to seven separate determinations of $K$. Each is subject to some experimental error. If the errors are randomly distributed, the best value of $K$ is given by the arithmetic mean

$$\bar{K} = \Sigma K_i/n$$

$n$ being the number of measurements, in this case seven.

An estimate of the correctness of the mean may be given in terms of the deviation of individual values from the mean. A standard deviation, $S$, is calculated as

$$S = [\Sigma(K_i - \bar{K})^2/(n - 1)]^{1/2}$$

For very large numbers of observations the standard deviation may be linked with exactness to the probability that an individual measurement will deviate from the mean by any given amount. For small numbers of observations, however, the standard deviations calculated in this way may themselves be subject to error, more particularly if the errors are not randomly distributed about the mean. The likely distribution of error in small sets of measurements was worked out by the statistician W. H. Gossett, writing under the name of "Student". He calculated so-called $t$-values by which $S$ should be multiplied to give the probability that a given measurement lies between stated limits. The $t$-values for five to seven measurements for various confidence levels are listed below.

### Table of *t*-values

| | Confidence level | | | | | | |
|---|---|---|---|---|---|---|---|
| $n$ | 50 | 60 | 70 | 80 | 90 | 95 | 99 |
| 5 | 0.74 | 0.94 | 1.19 | 1.53 | 2.13 | 2.78 | 4.60 |
| 6 | 0.73 | 0.92 | 1.16 | 1.48 | 2.02 | 2.57 | 4.03 |
| 7 | 0.72 | 0.91 | 1.13 | 1.44 | 1.94 | 2.45 | 3.71 |

To illustrate the use of these values, suppose we had six measurements of $K$ yielding $K = 270$ and $S = 17$. Suppose also that we wish a 95% chance of a new measurement lying within the stated range. We would then express our result thus far as

$$\bar{K} = 270 \pm t_{95} \times S$$
$$= 270 \pm 44.$$

Note that $t$ decreases as the number of measurements increases at a given confidence level and increases as the confidence limits are raised.

It may be that one or more of your measurements differs from the others in a nonrandom way. A simple check is provided by another statistical test, the so-called Q-test, which compares the spread of the values obtained with the deviations between the extreme values and their nearest neighbors.

Carry out the test as follows. Determine the difference between the largest and smallest values of $K$, the range of the measurements, and multiply this value by the Q-factor corresponding to the number of measurement. The Q-factors given below match the 90% probability level sought.

| $n$ | 3 | 4 | 5 | 6 | 7 | 8 |
|---|---|---|---|---|---|---|
| $Q_{90}$ | 0.90 | 0.70 | 0.58 | 0.50 | 0.45 | 0.41 |

Now calculate the absolute value of the difference between the largest and next largest values of $K$ and if this is greater than the Q × range product obtained above, reject the largest $K$. Carry out the same process on the smallest and next smallest values of $K$. Repeat the test on the remaining numbers, adjusting Q appropriately, until no further rejections occur.

Calculate $K$ and $S$ adjusted to the 90% confidence level from the values surviving the Q-test. If more than two values were rejected by the Q-test, ignore it and include all the values in your averaging and calculation of $S$.

Name _____ Instructor _____

## PRELABORATORY PREPARATION SHEET

1. Write an expression for the equilibrium constant for the following reaction taking place in aqueous solution:

$$3S^{2-}(aq) + 2MnO_4^-(aq) + 4H_2O \rightleftharpoons 3S(s) + 2MnO_2(s) + 8OH^-(aq)$$

2. Why does a large excess of $Fe^{3+}$ inhibit formation of complex ions containing more than one $SCN^-$ ligand?

3. What is the difference between absorption and absorbance?

4. Why do we not make an independent measurement of the molar absorptivity in this experiment?

5. The absorbance of a $0.010M$ solution of a complex ion is found to be 0.30 at 410 nm. What is the concentration of a second sample of this solution whose absorbance at the same wavelength is 0.40?

267

Name _____  Instructor _____

## DATA SHEET

1. Baseline absorbance of sample cell + distilled water:

   _____ (450 nm)

2. Absorbance of standard $FeSCN^{2+}$ concentration:

   (10 mL 200m$M$ $Fe(NO_3)_3$ + 10 mL 0.4m$M$ KSCN)

   _____

3. Absorbance of mixtures (concentrations before mixing, m$M$):

| $[Fe^{3+}]$ | $[SCN^-]$ | Absorbance | Baseline | Final $A$ |
|---|---|---|---|---|
| 20 | 0.40 | _____ | _____ | _____ |
| 20 | 0.20 | _____ | _____ | _____ |
| 10 | 0.40 | _____ | _____ | _____ |
| 10 | 0.20 | _____ | _____ | _____ |
| 5  | 0.40 | _____ | _____ | _____ |
| _____ | _____ | _____ | _____ | _____ |
| _____ | _____ | _____ | _____ | _____ |

Name _____  Instructor _____

## DATA SHEET

**1.** Baseline absorbance of sample cell + distilled water:

_____ (450 nm)

**2.** Absorbance of standard $FeSCN^{2+}$ concentration:

        (10 mL 200m$M$ $Fe(NO_3)_3$ + 10 mL 0.4m$M$ KSCN)

_____

**3.** Absorbance of mixtures (concentrations before mixing, m$M$):

| $[Fe^{3+}]$ | $[SCN^-]$ | Absorbance | Baseline | Final $A$ |
|---|---|---|---|---|
| 20 | 0.40 | _____ | _____ | _____ |
| 20 | 0.20 | _____ | _____ | _____ |
| 10 | 0.40 | _____ | _____ | _____ |
| 10 | 0.20 | _____ | _____ | _____ |
| 5 | 0.40 | _____ | _____ | _____ |
| _____ | _____ | _____ | _____ | _____ |
| _____ | _____ | _____ | _____ | _____ |

Name _____  Instructor _____

# REPORT SHEET

**1. Standard FeSCN$^{2+}$ concentration and $Eb$:**

| Before mixing: | | After mixing: | | At equilibrium: | | | |
|---|---|---|---|---|---|---|---|
| [Fe$^{3+}$] | [SCN$^-$] | [Fe$^{3+}$] | [SCN$^-$] | [FeSCN$^{2+}$] | [Fe$^{3+}$] | [SCN$^-$] | $A_{450}$ |
| _____ | _____ | _____ | _____ | _____ | _____ | _____ | _____ |

Experimental absorptivity $Eb$ _____

**2. Mixture concentrations and $K$:**

| Before mixing: | | After mixing: | | |
|---|---|---|---|---|
| [Fe$^{3+}$] | [SCN$^-$] | [Fe$^{3+}$] | [SCN$^-$] | $A_{450}$ |
| _____ | _____ | _____ | _____ | _____ |
| _____ | _____ | _____ | _____ | _____ |
| _____ | _____ | _____ | _____ | _____ |
| _____ | _____ | _____ | _____ | _____ |
| _____ | _____ | _____ | _____ | _____ |
| _____ | _____ | _____ | _____ | _____ |
| _____ | _____ | _____ | _____ | _____ |

| At equilibrium | | | |
|---|---|---|---|
| [FeSCN$^{2+}$] | [Fe$^{3+}$] | [SCN$^-$] | $K$ |
| _____ | _____ | _____ | _____ |
| _____ | _____ | _____ | _____ |
| _____ | _____ | _____ | _____ |
| _____ | _____ | _____ | _____ |
| _____ | _____ | _____ | _____ |
| _____ | _____ | _____ | _____ |
| _____ | _____ | _____ | _____ |

**3. Mean $K$ and standard deviation:**

$K =$ _____   $S =$ _____

$S \cdot t_{90} =$ _____

Number of values used in calculation _____

## WORKSHEET

Carry out all the calculations for Experiment 13 on this sheet, attach it to your report sheet, and submit the two together. Do your calculations directly on this sheet. Do not do them on scratch paper and then copy them on to this sheet. Make no erasures and delete incorrect entries with a single line so as to leave the original data legible. Use a pen.

# EXPERIMENT 14

# Acid-Base Equilibria Part I. The Use of Indicators to Determine pH

### AIMS

To determine the useful range of certain indicators using solutions of known pH, to use these indicators to determine the pH of various acid and base solutions, and to find acid and base dissociation constants for these solutions from their measured pH.

### EQUIPMENT

From locker: 20 test tubes
25-mL graduated cylinder
Four 30-mL beakers
Laboratory: Test tube stand

### REAGENTS AND OTHER MATERIALS

$1M$, $0.1M$, $0.01M$ HCl
$0.01M$ NaOH
$0.1M$ acetic acid

0.1$M$ sodium acetate
0.1$M$ ammonia
0.1$M$ ammonium chloride
Buffer solutions of pH 3, 4, 5, 6, 7, 8, 9 10, and 11
Indicators: Methyl orange, methyl red, methyl violet, phenolphthalein, bromthymol blue, and alizarin yellow R
Wax pencil for marking test tubes

## BACKGROUND

### Associated Readings

The following sections of *Chemical Principles* should be read in conjunction with Experiment 14.

**Chapter 15** Acids and Bases

All of this important chapter with the exception of Sections 15.3 and 15.8 should be carefully studied in preparation for Experiment 14 and reviewed before tackling Experiments 16 and 17.

---

Acids are classified as **strong** or **weak** depending on the extent to which they react with water to yield hydrated protons.

$$HA(aq) + H_2O \rightleftharpoons H_3O^+(aq) + A^-(aq)$$

To emphasize the proton-transfer aspect of this reaction we write the hydrated proton as $H_3O^+(aq)$. The proton in aqueous solution is a highly mobile species and $H_3O^+$ is only one of several forms of hydrated proton present. In later sections we will use the simplified form

$$HA(aq) \rightleftharpoons H^+(aq) + A^-(aq)$$

to represent the reaction between an acid and water, using the symbol $H^+(aq)$ to represent the hydrated proton.

With **strong acids** the reaction proceeds effectively to completion and the concentration of hydrated protons formed is equal to the formal concentration of the acid itself. Typical strong acids include $HCl$, $HNO_3$, $H_2SO_4$, $HClO_4$, etc.

In **weak acids** the reaction with water is incomplete. Typical weak acids include $HF$, $HCN$, and organic carboxylic acids such as formic acid, $HCOOH$, acetic acid, $CH_3COOH$, and benzoic acid, $C_6H_5COOH$, as well as hydroxylated aromatic compounds such as phenol, $C_6H_5OH$. Aromatic compounds are organic compounds incorporating one or more benzene rings in their structure. (See Sections 21.4 and 21.5 of *Chemical Principles*.)

A measure of the acid strength of a weak acid is afforded by the equilibrium constant for the reaction of the acid with water. Representing the weak acid as HA, then

$$HA(aq) \rightleftharpoons H^+(aq) + A^-(aq)$$

the constant for this reaction is normally expressed as the **acid dissociation constant**,

$$K_a = \frac{[H^+][A^-]}{[HA]}$$

The smaller the value of $K_a$, the weaker the acid. For acetic acid, at room temperature, $K_a$ is $2 \times 10^{-5}$. For HCN, $K_a$ is $5 \times 10^{-10}$. Acetic acid is thus a stronger acid than hydrocyanic acid, though both are classified as weak acids.

If $K_a$ is known for a particular acid then we can calculate the concentration of hydrated protons in an aqueous solution of the acid of known formal concentration. Consider as an example a 0.040$M$ solution of HCOOH for which $K_a$ is $1.7 \times 10^{-4}$. We have

$$HCOOH(aq) \rightleftharpoons HCOO^-(aq) + H^+(aq)$$

and

$$K_a = \frac{[HCOO^-][H^+]}{[HCOOH]}$$

Suppose the concentration of hydrated protons at equilibrium, $[H^+]$, to be $x\ M$, then since $[HCOO^-] = [H^+]$ by stoichiometry, and since [HCOOH] will then be $(0.040 - x)\ M$, we may write

$$\frac{x^2}{0.040 - x} = 1.7 \times 10^{-4}$$

Solution of the resultant quadratic yields $x = 2.5 \times 10^{-3} M$.

We can simplify the calculation if we assume that $x$ is very small with respect to the formal concentration of unreacted acid. This assumption is usually justified for dilute solutions of acids whose $K_a$ is less than about $10^{-4}$. We therefore write

$$\frac{x^2}{0.040 - x} \simeq \frac{x^2}{0.040} = 1.7 \times 10^{-4}$$

The resultant value of $x$, $2.6 \times 10^{-3}$, differs from the value found by solution of the quadratic by only 4%, an acceptable level for most purposes. As a general rule one should check the hydrated proton concentration obtained by solution of the simplified form against the formal concentration of the acid to make certain the assumption as to their relative magnitudes is justified.

**p Operators,** such as **pH,** use a logarithmic scale to express quantities whose size is very small but whose range may be large, covering many orders of magnitude.

In general terms

$$pX = -\log_{10} X$$

pH as a measure of hydrated proton concentration is defined as

$$pH = -\log_{10}[H^+(aq)]$$

A solution of 0.1$M$ HCl, which is completely dissociated, has

$$[H^+(aq)] = 1 \times 10^{-1}\ M,$$

so that

$$pH = -\log_{10}(1 \times 10^{-1})$$
$$= -(-1)$$
$$= 1$$

For a 0.01$M$ solution of HCl, pH would be 2; for a 0.001$M$ solution, pH would be 3, and so on. A single unit change in pH therefore reflects a tenfold change in hydrated proton concentration.

For a weak acid pH cannot be determined in this simple way from the formal molar concentration. A separate calculation must first be carried out to find $[H^+]$. For the 0.040$M$ solution of formic acid considered above

$$pH = -\log_{10}(2.5 \times 10^{-3})$$
$$= -(\log_{10} 2.5 + \log_{10} 10^{-3})$$
$$= -(0.4 - 3.000)$$
$$= -(-2.6)$$
$$= 2.6$$

**Acidity** and **basicity,** the terms used to describe the condition of aqueous solutions which contain acids and bases, are determined with respect to pure water in whose autoionization equal amounts of $H^+$ and $OH^-$ are formed.

For the reaction

$$H_2O \rightleftharpoons H^+(aq) + OH^-(aq)$$

the equilibrium constant is written as

$$K_w = [H^+][OH^-]$$

and has, for pure water at room temperature, the value $1.0 \times 10^{-14}$. The equilibrium concentration of hydrated protons in pure water is thus $1.0 \times 10^{-7}M$, equal to the concentration of hydroxide ions. The pH of pure water is thus 7. Water having this pH is defined as **neutral.** Note that natural waters generally have a pH lower than 7 because of $CO_2$ absorption:

$$CO_2(g) + H_2O \rightleftharpoons HCO_3^-(aq) + H^+(aq)$$

In an aqueous acidic solution the hydrated proton concentration is greater than that of hydroxide ion and the pH of such a solution is less than 7. In a basic solution the hydroxide ion concentration is greater than that of hydrated protons; $[H^+]$ will be less than $10^{-7}M$, and pH will be greater than 7.

The concentrations of hydrated protons and hydroxide ions in any aqueous solution are obviously dependent on one another, as defined by $K_w$; the product of their equilibrium concentrations is always equal to $10^{-14}$ at 25°C. Thus, in $0.10M$ HCl solution where $[H^+]$ is $1.0 \times 10^{-1}$, $[OH^-] = K_w/[H^+] = 1.0 \times 10^{-13}$ and pOH = 13. The relationship

$$pH + pOH = pK_w = 14$$

is apparent.

Among the most common strong bases are the hydroxides of the Group IA metals and of calcium. Weak bases include ammonia and organic amines, compounds containing N atoms capable of accepting a proton.

The $K_b$ for a solution of a weak base of known concentration and pH is found in exactly the same way as $K_a$ for a weak acid. For a base B in water

$$H_2O + B(aq) \rightleftharpoons HB^+(aq) + OH^-(aq)$$

$$K_b = \frac{[OH^-][HB^+]}{[B]}$$

and $[OH^-]$ is found from pH using the relation $pH + pOH = pK_w$.

The conjugate species $HB^+$ is itself a weak acid for which $K_a$ can be found in the usual way:

$$HB^+(aq) \rightleftharpoons H^+(aq) + B(aq)$$

$$K_a = \frac{[H^+][B]}{[HB^+]}$$

For a conjugate acid-base pair in solution the product of their dissociation constants,

$$K_a \cdot K_b = \frac{[H^+][B]}{[HB^+]} \times \frac{[OH^-][HB^+]}{[B]} = [H^+][OH^-]$$

reduces to $K_w$. We can find $K_b$ for the base from $K_a$ for the conjugate acid, so that tables of dissociation constants are usually given in terms of $K_a$ for the neutral or cationic species.

## INDICATORS

One of the simplest methods of measuring pH is to make use of chemical **indicators.** These are either synthetic or naturally occurring organic compounds of intermediate molecular weight which differ in color depending on the pH of the solution in which

## Acid-Base Equilibria Part I. The Use of Indicators to Determine pH

they are dissolved. One of the oldest indicators is litmus, whose red color in acid solution and blue in base are well known. A similar color change occurs with grape juice.

Chemically, indicators are generally compounds which act as weak acids or weak bases in aqueous solution and therefore take part in equilibria involving hydrated protons. If we signify an indicator which is a weak acid by HIn, then its reaction with water may be written

$$HIn(aq) \rightleftharpoons H^+(aq) + In^-(aq)$$

To be useful the colors of dissolved undissociated HIn and the anion In$^-$ must be different. In a strongly acidic solution, with a preponderance of H$^+$ present, Le Chatelier's principle dictates that the equilibrium in the above reaction will favor the reactants and so the color of the solution will match that of the undissociated acid form of the indicator. By similar reasoning the color of a strongly basic solution will be that of In$^-$.

This does not mean that *all* indicators will show one color in acid solution and another in basic solution. The pH at which the color change takes place depends on the acid or base strength of the indicator itself.

$K_a$ for the indicator dissociation is written as

$$\frac{[H^+][In^-]}{[HIn]}$$

which may be rearranged to give

$$\frac{[In^-]}{[HIn]} = \frac{K_a}{[H^+]}$$

With variation in the pH of the solution in which the indicator has been dissolved there will be a variation in the ratio [In$^-$]/[HIn] and hence a change in the color of the solution, which reflects the relative amounts of the two species present. The pure color of either species will generally be perceived by the human eye when that species is present in about tenfold excess over the other and intermediate colors will be noted in the intervening range of concentration. The color of the undissociated form of the indicator is detected for solutions in which

$$[In^-]/[HIn] \leq 1/10$$

and the color of the dissociated form in solutions for which

$$[In^-]/[HIn] \geq 10/1.$$

Substituting these numerical ratios into the expression for $K_a$ for the indicator shows that for the color of the undissociated form to show clearly

$$K_a = \frac{[H^+][In^-]}{[HIn]} = \frac{[H^+] \times 1}{10}$$

or
$$[H^+] = 10 K_a$$

For the color of the dissociated form to show clearly

$$K_a = \frac{[H^+] \times 10}{1}$$

or
$$[H^+] = K_a/10$$

Using p functions: the color of the undissociated form shows when

$$pH = -1 - \log K_a = pK_a - 1$$

the color of the dissociated form shows when

$$pH = +1 - \log K_a = pK_a + 1$$

**TABLE 14.1** Useful Ranges of Some Indicators

| Indicator | pH Range |
|---|---|
| Methyl violet | 0.2 to 1.6 |
| Methyl orange | 3.2   4.4 |
| Methyl red | 4.8   6.0 |
| Bromthymol blue | 6.0   7.6 |
| Phenol red | 6.4   8.0 |
| Phenolphthalein | 8.0   10.0 |
| Alizarin yellow | 10.1   12.0 |

Clear indications of different colors are therefore to be expected within one pH unit on either side of a pH equal to the pK value for the indicator. These different colors are often referred to as the "acid" and "base" colors of the indicators, but one must be careful not to confuse these terms with the colors exhibited by the indicator in acidic or basic solution.

As an example, consider the indicator methyl orange for which $K_a$ is $1.6 \times 10^{-4}$ and pK 3.8. The red color of the undissociated indicator (its acid form) would be seen in solutions of pH 2.8 or less, the yellow color of the indicator anion (its base form) would be seen in solutions of pH 4.8 and above. Intermediate orange colors would be noted in the pH range between these two values, the so-called **transition range.** Because of the greater sensitivity of the eye in the orange region of the visible spectrum, the transition range for this indicator is actually somewhat sharper, being from 3.1 to 4.4 in pH. Note, however, that both solutions are acidic!

We cannot completely describe the pH of a solution simply by observing the color of the acid or base form of the indicator. With methyl orange, if the "acid" color is seen in a particular solution all that can be said is that the pH is 2.8 or less; if the "base" color is seen all that can be said is that the pH is 4.8 or more. However, by experimenting with a series of indicators whose pK values span the usual pH range we can narrow our estimate of the pH of any given solution to within acceptable limits.

The transition ranges of the indicators to be used in this experiment are given in Table 14.1. Where the indicator is used to determine a pH specifically by observation of the intermediate color, the transition range is often referred to as the **useful range** of the indicator. In this experiment we determine the characteristic colors of the acid and base forms of the indicators and use them to establish the pH of solutions of acetic acid and sodium acetate, ammonia and ammonium chloride. From the measured pH we independently calculate $K_a$ and $K_b$ for the two conjugate acid-base pairs and confirm the relationship $K_a \cdot K_b = K_w$.

## PROCEDURE

### SAFETY NOTES
You are working today with solutions which span a wide range of acid-base concentrations. Be properly careful in handling these materials.

The experiment is carried out by students working in pairs. You will need to pool your glassware, so be sure to identify items belonging to each member of the team. All glassware must be scrupulously clean before beginning the experiment. (Recall the effect on the pH of water of small amounts of acid or base.)

### 1. Standardization of Indicators

Wash and clean 18 test tubes, rinsing them with distilled water. Number them consecutively 1–18 with a wax pencil. Add 10 mL of distilled water to another test tube with a graduated cylinder. Using this test tube as a guide, add 10 mL of each of the reference pH solutions listed in Part I of your data sheet to the remaining test tubes in order. Support the test tubes in a test tube rack. To each solution add 1 drop (*and only one drop*) of the appropriate indicator as listed on the data sheet and mix the indicator and the solution by placing your *clean, dry* thumb over the mouth of the tube and repeatedly inverting the tube. Record the color shown by the indicator in each case. To determine the colors it is helpful to sight down the long axis of the tube and to view the color against a white background. Be reasonably concise in your descriptions of the colors, but recognize effective gradations in color. Thus, pink and magenta can be distinguished from vermilion and red. Save the 18 numbered solutions for later reference.

## 2. Determination of the pH of the 0.1*M* Solutions

In clean, dry labeled 30-mL beakers collect 25 mL each of the 0.1*M* acetic acid, sodium acetate, ammonia, and ammonium chloride solutions. On a watchglass test small amounts of each solution in turn against the various indicators, using very small drops of indicator until you have settled on the best choice to determine the pH. When you have decided on an indicator (or indicators, a better estimate can often be made by using two indicators to bracket the pH) for a given solution, then add 10 mL of that solution to a test tube and add one drop of the indicator, mix the solution and compare its color with that of the reference pH solutions for that indicator. Use the color matches to establish the pH of the four solutions as accurately as possible.

Name _____ Instructor _____

## PRELABORATORY PREPARATION SHEET

1. Write a balanced equation for the reaction of the weak acid hydrogen cyanide, HCN, with water.

   _____

   The numerical value of $K_a$ for HCN is $5 \times 10^{-10}$. Write an expression for $K_a$ for this acid and find the pH of a $0.01M$ solution of HCN.

   $K_a =$ _____  (give expression)

   _____
   _____
   _____
   _____

   $[H_3O^+] =$ _____

   pH = _____

2. Write a balanced net ionic equation for the reaction of $NH_4NO_3$ with water

   _____

   Give an expression for $K_a$ for $NH_4^+$

   $K_a =$

3. If $K_b$ for ammonia is $2 \times 10^{-5}$, what is the numerical value of $K_a$ for ammonium ion? Show your reasoning.

   _____
   _____

4. Why is it important to use only small amounts of indicator in determining the pH of a solution?

   _____
   _____

285

Name _____  Instructor _____

Partner _____

## DATA SHEET

Eighteen solutions are listed below together with the suggested indicator for test purposes. Prepare the mixtures as directed in Section 1 of the procedure and record the observed color.

| Test Solution | Indicator | Color |
|---|---|---|
| 1. $1M$ HCl | Methyl violet | _____ |
| 2. $0.1M$ HCl | Methyl violet | _____ |
| 3. $0.01M$ HCl | Methyl violet | _____ |
| 4. pH 3 buffer | Methyl orange | _____ |
| 5. pH 4 buffer | Methyl orange | _____ |
| 6. pH 5 buffer | Methyl orange | _____ |
| 7. pH 4 buffer | Methyl red | _____ |
| 8. pH 5 buffer | Methyl red | _____ |
| 9. ph 6 buffer | Methyl red | _____ |
| 10. pH 6 buffer | Bromthymol blue | _____ |
| 11. pH 7 buffer | Bromthymol blue | _____ |
| 12. pH 8 buffer | Bromthymol blue | _____ |
| 13. pH 8 buffer | Phenol phthalein | _____ |
| 14. pH 9 buffer | Phenol phthalein | _____ |
| 15. pH 10 buffer | Phenol phthalein | _____ |
| 16. pH 10 buffer | Alizarin yellow | _____ |
| 17. pH 11 buffer | Alizarin yellow | _____ |
| 18. $0.01M$ NaOH | Alizarin yellow | _____ |

Be sure that the test tubes are clearly numbered for easy identification.

## Determination of pH

$0.1M$ acetic acid         pH = _____ ± _____

Indicator(s) _____

Name _____  Instructor _____

Partner _____

## DATA SHEET

Eighteen solutions are listed below together with the suggested indicator for test purposes. Prepare the mixtures as directed in Section 1 of the procedure and record the observed color.

| Test Solution | Indicator | Color |
|---|---|---|
| 1. 1$M$ HCl | Methyl violet | _____ |
| 2. 0.1$M$ HCl | Methyl violet | _____ |
| 3. 0.01$M$ HCl | Methyl violet | _____ |
| 4. pH 3 buffer | Methyl orange | _____ |
| 5. pH 4 buffer | Methyl orange | _____ |
| 6. pH 5 buffer | Methyl orange | _____ |
| 7. pH 4 buffer | Methyl red | _____ |
| 8. pH 5 buffer | Methyl red | _____ |
| 9. ph 6 buffer | Methyl red | _____ |
| 10. pH 6 buffer | Bromthymol blue | _____ |
| 11. pH 7 buffer | Bromthymol blue | _____ |
| 12. pH 8 buffer | Bromthymol blue | _____ |
| 13. pH 8 buffer | Phenol phthalein | _____ |
| 14. pH 9 buffer | Phenol phthalein | _____ |
| 15. pH 10 buffer | Phenol phthalein | _____ |
| 16. pH 10 buffer | Alizarin yellow | _____ |
| 17. pH 11 buffer | Alizarin yellow | _____ |
| 18. 0.01$M$ NaOH | Alizarin yellow | _____ |

Be sure that the test tubes are clearly numbered for easy identification.

### Determination of pH

0.1$M$ acetic acid        pH = _____ ± _____

Indicator(s) _____

0.1*M* sodium acetate       pH = _____ ± _____

   Indicator(s) _____

0.1*M* ammonia              pH = _____ ± _____

   Indicator(s) _____

0.1*M* ammonium chloride    pH = _____ ± _____

   Indicator(s) _____

0.1$M$ sodium acetate        pH = _____ ± _____

   Indicator(s) _____

0.1$M$ ammonia              pH = _____ ± _____

   Indicator(s) _____

0.1$M$ ammonium chloride    pH = _____ ± _____

   Indicator(s) _____

Name _____  Instructor _____

## REPORT SHEET

**1.** From your data sheet enter the acid and base colors of the listed indicators.

| Indicator | Acid form | Base form |
|---|---|---|
| Methyl violet | | |
| Methyl orange | | |
| Methyl red | | |
| Bromthymol blue | | |
| Phenolphthalein | | |
| Alizarin yellow | | |

**2.** From the observed pH, calculate p$K$ for each solution tested. Given the equation for the reactions of each species with water, list the concentrations of all relevant species, give the expression for $K$ as well as its numerical value, and estimate the uncertainty in p$K$.

### 0.1$M$ Acetic acid, $CH_3COOH$:

Equation _____

pH = _____ ± _____   [$H^+(aq)$] _____

Concentrations of other relevant species _____

_____

$K_a$ =

p$K_a$ = _____ ± _____

### 0.1$M$ Sodium acetate, $CH_3COONa$:

Equation _____

pH = _____ ± _____   [$H^+(aq)$] _____

Concentrations of other relevant species _____

_____

$K_b$ =

p$K_b$ = _____ ± _____

**0.1 M Ammonia, $NH_3$:**

Equation _____

pH = _____ ± _____   [$H^+$(aq)] _____

Concentrations of other relevant species _____

_____


$pK_b$ = _____ ± _____


**0.1 M Ammonium chloride, $NH_4Cl$:**

Equation _____

pH = _____ ± _____   [$H^+$(aq)] _____

Concentrations of other relevant species _____

_____


$K_a$ =

$pK_a$ = _____ ± _____

**3.** For each conjugate acid-base pair, calculate $pK_a + pK_b$.

Acetic acid–acetate ion _____ ± _____

Ammonia–ammonium ion _____ ± _____

297

## WORKSHEET

Carry out all the calculations for Experiment 14(I) on this sheet, attach it to your report sheet, and submit the two together. Do your calculations directly on this sheet. Do not do them on scratch paper and then copy them on to this sheet. Make no erasures and delete incorrect entries with a single line so as to leave the original data legible. Use a pen.

# EXPERIMENT 14

# Acid-Base Equilibria Part II. Buffer Solutions and the pH Meter

### AIMS

To prepare a buffer solution, to examine its behavior in response to dilution and to added acid and base, and to learn the use of the pH meter.

### EQUIPMENT

From stockroom: 5-mL graduated cylinder
From locker: 50-mL beakers (3)
100-mL beakers (2)
Stirring rod
25-mL graduated cylinder
Laboratory: pH meter

### REAGENTS AND OTHER MATERIALS

$0.10M$ acetic acid
$0.10M$ sodium acetate

0.10$M$ ammonia
0.10$M$ ammonium chloride
0.10$M$ HNO$_3$ (bulk and in dropper bottles)
0.10$M$ NaOH (bulk and in dropper bottles)

## BACKGROUND

If you have not yet carried out Part I of this experiment, you should first read through the Background sections in that part that deal with acids, bases, and p-operators.

When the salt of a weak acid such as sodium formate dissolves in water, the pH of the resultant solution is basic. Correspondingly, the pH of an aqueous solution of the salt of a weak base, such as ammonium chloride, is acidic. These observations are readily understandable in Brønsted terms. Formate anion, the anion present in sodium formate, is the conjugate base of the weak acid formic acid, and an aqueous solution reacts with water acting as an acid, to produce a small amount of undissociated formic acid and an equal amount of OH$^-$ ion.

$$HCOO^-(aq) + H_2O \rightleftharpoons HCOOH(aq) + OH^-(aq)$$

[OH$^-$] is then greater than the $1.0 \times 10^{-7}M$ concentration of the species present in neutral water, so the solution is basic with pH > 7. The exact concentration of OH$^-$ may be found from the measured pH of the solution, using the relationship

$$pH + pOH = pK_w$$

or from the known value of $K_a$ for the conjugate acid, using the relationship

$$K_a \cdot K_b = K_w$$

Correspondingly, ammonium ion is the conjugate acid of the weak base ammonia, and an aqueous solution reacts with water acting as a base, to yield a solution in which [H$^+$(aq)] = $1.0 \times 10^{-7}M$, pH < 7.

$$NH_4^+(aq) + H_2O \rightleftharpoons NH_3(aq) + H_3O^+(aq)$$

The pH of pure water itself is 7, reflecting the hydrated proton concentration of $1 \times 10^{-7}$. However, water shows a marked instability in pH with respect to addition of only small amounts of acid or base. Addition of only 0.1 mL (two drops) of 1$M$ nitric acid to one liter of water is enough to change the pH from 7 to 4, a thousandfold change in hydronium ion concentration. In the same way addition of only two drops of 1.0$M$ sodium hydroxide to a liter of neutral water would change its pH to 10, a thousandfold increase in hydroxide ion concentration over that in neutral water.

In biological systems where all the metabolic reactions take place in aqueous solution, it is critically important that pH remains stable, as many biological macromolecules function only within very narrow limits of pH and many are destroyed by extremes in pH. How then is pH stability maintained in these aqueous systems, given the extreme instability of water in response to very small additions of acid or base? It is done through a simple yet subtle mechanism which is dependent on the behavior of weak acids or weak bases and their soluble salts in solutions in which both are present.

The principles are illustrated in this example. Consider a solution which is 0.20$M$ with respect to formic acid and 0.16$M$ with respect to sodium formate. In

$$HCOOH(aq) \rightleftharpoons H^+(aq) + HCOO^-(aq)$$

for which $K_a = 1.7 \times 10^{-4}$, we may substitute the approximations [HCOOH] = 0.20$M$ and [HCOO$^-$] = 0.16$M$, yielding [H$^+$] = $2.2 \times 10^{-4}M$ and pH = 3.7.

Consider the effect on pH of adding 5.0 mL of $1.0M$ $HNO_3$ to 200 mL of this solution. Note that on a relative basis we are now adding 250 times the amount of acid we added to the water sample.

Addition of a strong acid causes production of an equivalent amount of formic acid through the reaction

$$HCOO^-(aq) + H^+(aq) \rightleftharpoons HCOOH(aq) + H_2O$$

for which the equilibrium lies well to the right. As the concentration of formic acid increases there is a stoichiometrically equivalent decrease in the concentration of formate anion. Let us calculate the new concentrations of these two species and again find the pH of the solution.

$$[HCOOH] = [(0.200 \text{ L} \times 0.20M) + (0.0050 \text{ L} \times 1M)]/0.205 \text{ L}$$
$$= 0.22M$$
$$[HCOO^-] = [0.200 \text{ L} \times 0.16M) - (0.0050 \times 1M)]/0.205 \text{ L}$$
$$= 0.13M$$

By substitution we find $[H^+] = 2.9 \times 10^{-4}$ and pH = 3.5.

We see that in this solution addition of a substantial amount of acid has had little effect on the pH, the hydrated proton concentration having increased by only 40% rather than by a factor of more than $10^5$, which would have been the case had the same amount of acid been added to 200 mL of neutral water.

An analogous calculation for the effect of added base shows the same resistance to change in pH, the **buffering action** in this case resulting from the reaction

$$OH^-(aq) + HCOOH(aq) \rightleftharpoons HCOO^-(aq) + H_2O$$

The amount of added acid that can be neutralized by a buffer solution depends on the amount of buffer base present, just as the amount of added base that can be neutralized depends on the amount of buffer acid present. Maximum buffering against either added acid or base results when the buffer acid and base are present in equal amounts and in fairly high concentrations. The pH of the buffer solution may be calculated from $K_a$ for the buffer acid according to the relation

$$[H^+(aq)] = K_a \frac{[HA]}{[A^-]}$$

which yields the Henderson–Hasselbach equation:

$$pH = pK_a + \log_{10}([A^-]/[HA])$$

where HA is the buffer acid and $A^-$ is its conjugate base.

Where $[HA] = [A^-]$, the equation reduces to

$$pH = pK_a$$

allowing us to choose an appropriate acid when a buffer of particular pH is needed.

In living cells the principal buffering effect is through equilibria involving carbonic acid, which results from the dissolution of carbon dioxide in water, and the bicarbonate anion $HCO_3^-$. $K_a$ for carbonic acid is $4.6 \times 10^{-7}$, so that near neutral pH is obtained.

The Henderson–Hasselbach equation gives a reasonable estimate of pH as long as the ratio of buffer acid to base is neither very large nor very small. In this experiment you prepare a buffer solution of a known concentration in a range where the equation works reasonably well and test its behavior against dilution and added acid and base. The Henderson–Hasselbach equation predicts that because dilution should affect the concentration of buffer acid and base equally, pH should remain relatively unchanged. You check this prediction and the prediction of buffer solution pH upon addition of acid or base.

## PROCEDURE

### SAFETY NOTES
You are working today with solutions which span a wide range of acid-base concentrations. Be properly careful in handling these materials.

The bulbs on the pH meter electrodes are quite fragile and you should be careful not to break them. The bulbs should not be allowed to dry out and should be immersed in distilled water when you have finished working with them.

The experiment is carried out by students working in pairs. You will need to pool your glassware, so be sure to identify items belonging to each member of the team. All glassware must be scrupulously clean before beginning the experiment. (Recall the effect on the pH of water of small amounts of acid or base.) Each pair of students should check out a 5-mL graduated cylinder from the stockroom. It is to be thoroughly cleaned before being returned.

### 1. Using the pH Meter

The pH meter is a sensitive delicate instrument by which the pH of a solution may be determined from the electrical conductivity due to hydrated protons. Please handle the instrument with care and according to the directions provided for the particular model in use in your laboratory. The electrode bulbs are of thin glass and are fragile and expensive; you may be charged for any breakage. If properly handled the instrument will give reliable results, if mishandled bad results are inevitable.

Because most instruments take some time to equilibrate, the meter will most likely have been switched on before the class period begins and the temperature control will have been set to the temperature of the laboratory. Do not alter this setting.

The meter is first calibrated by dipping the electrode into a pH 7 buffer and setting the dial at exactly that pH. We can find the pH of an unknown solution by first rinsing the electrode with distilled water, and then transferring it to the unknown solution and reading the dial on the meter. Specific instructions for calibration will be posted close to the meter. Always follow these instructions carefully.

The following procedures apply generally:

**a.** Before transferring the electrode from one solution to another, always set the meter to STANDBY.
**b.** Always thoroughly and carefully rinse the electrode by directing a stream of distilled water from your wash bottle around and over the bulb before inserting it into a new solution. Failure to remove all traces of the old solution may invalidate your readings. Use a beaker to collect the washings.
**c.** When you have finished with the meter leave the electrode bulb immersed in a beaker of distilled water. Such a beaker properly labeled, should be ready by each meter. If you cannot find it, inform your assistant at once. The electrode bulb must not be allowed to dry out under any circumstances or the system will fail.
**d.** Always leave the meter in the STANDBY mode when you have finished using it.

### 2. Determination of the pH of Distilled Water in Response to Added Acid or Base

To each of two clean 30-mL beakers add 20 mL of distilled water. Use the pH meter to find the pH of each sample. Recall that though the water is distilled it is not necessarily pure and its pH may not be exactly 7.

Now from the labeled dropper bottles add to one beaker two drops (0.1 mL) of 0.1$M$ HNO$_3$, and to the other two drops of 0.1$M$ NaOH. Measure and record the pH of each sample.

## 3. Determination of the pH of a Buffer Solution in Response to Added Acid or Base and Dilution

You will be assigned either an acetic acid–sodium acetate or an ammonia–ammonium chloride buffer system to examine. Instructions are given here for the acetic acid–acetate system but ingredients for the ammonia–ammonium system are given in parentheses. Follow the same procedures in either case. Be sure to rinse the bulb on the pH meter between each reading of pH.

In two separate clean, dry 100-mL beakers collect 35 mL each of 0.1$M$ acetic acid and 0.1$M$ sodium acetate (0.1$M$ ammonia, 0.1$M$ ammonium chloride).

Use these solutions to prepare three samples of a buffer solution containing 10 mL of each ingredient. Prepare the buffer solutions in clean, dry 50-mL beakers and use a graduated cylinder to measure the volumes as accurately as possible. Label the beakers A, B, and C.

Mix the contents of each beaker thoroughly with a stirring rod and record the pH of each. To test the effect of dilution on pH, add 5 mL of distilled water to solution A, mix thoroughly, and record the pH. Add a further 15 mL of distilled water in 5 mL increments, mixing and recording the pH after each addition.

To test the effect of added acid, add 15 mL of 0.1$M$ HNO$_3$ to solution B in 1-mL increments (use the 5-mL graduated cylinder). Mix the contents of the beaker thoroughly after each addition and record the pH at each stage.

To test the effect of added base, add 15 mL of 0.1$M$ NaOH to the solution in beaker C, again in 1-mL increments, mixing thoroughly and recording the pH after each addition.

When you have completed the measurements, discard the various solutions, flushing them down the sink with cold water. Carefully rinse off the bulb of the pH meter and replace it in the distilled water container provided.

Name _____  Instructor _____

## PRELABORATORY PREPARATION SHEET

1. Define what is meant by the term "buffer solution" (a) in functional terms, (b) in terms of its chemical composition.

   (a) _____

   _____

   (b) _____

   _____

2. Find the pH of a buffer solution which is $0.25M$ in sodium acetate and $0.30M$ in acetic acid, if $K_a$ for acetic acid at 25°C is taken as $1.0 \times 10^{-5}$. Show the steps in your calculation.

   _____

   _____

   _____

   _____

3. What is the pH of a buffer solution in which the concentration of acetic acid and potassium acetate are stoichiometrically equal.

   _____

4. Benzoic acid, $C_6H_5COOH$, has $K_a = 6.5 \times 10^{-5}$ at 25°C. What masses of benzoic acid and sodium benzoate, $C_6H_5COONa$, should be taken to prepare 250 mL of a buffer solution of pH 4.0 if the sum of the molar concentrations of the two species is not to exceed $0.50M$?

   _____

   _____

   _____

   _____

Name _____    Instructor _____

Partner _____

## DATA SHEET

**1.** Distilled Water Samples:

|  | pH |  | pH |
|---|---|---|---|
| A alone | _____ | B alone | _____ |
| A + 0.1 mL 0.1$M$ HNO$_3$ | _____ | B + 0.1 mL 0.1$M$ NaOH | _____ |

**2.** Buffer Samples:

|  | pH |
|---|---|
| A alone | _____ |
| A + 5 mL H$_2$O | _____ |
| A + 10 mL H$_2$O | _____ |
| A + 15 mL H$_2$O | _____ |
| A + 20 mL H$_2$O | _____ |

| Added 0.1$M$ HNO$_3$ (mL) | pH | Added 0.1$M$ NaOH (mL) | pH |
|---|---|---|---|
| B   0. | _____ | C   0. | _____ |
| 1. | _____ | 1. | _____ |
| 2. | _____ | 2. | _____ |
| 3. | _____ | 3. | _____ |
| 4. | _____ | 4. | _____ |
| 5. | _____ | 5. | _____ |
| 6. | _____ | 6. | _____ |
| 7. | _____ | 7. | _____ |
| 8. | _____ | 8. | _____ |
| 9. | _____ | 9. | _____ |
| 10. | _____ | 10. | _____ |
| 11. | _____ | 11. | _____ |
| 12. | _____ | 12. | _____ |
| 13. | _____ | 13. | _____ |

Name _____   Instructor _____

Partner _____

## DATA SHEET

**1.** Distilled Water Samples:

|  | pH |  | pH |
|---|---|---|---|
| A alone | _____ | B alone | _____ |
| A + 0.1 mL 0.1$M$ HNO$_3$ | _____ | B + 0.1 mL 0.1$M$ NaOH | _____ |

**2.** Buffer Samples:

|  | pH |
|---|---|
| A alone | _____ |
| A + 5 mL H$_2$O | _____ |
| A + 10 mL H$_2$O | _____ |
| A + 15 mL H$_2$O | _____ |
| A + 20 mL H$_2$O | _____ |

| Added 0.1$M$ HNO$_3$ (mL) | pH | Added 0.1$M$ NaOH (mL) | pH |
|---|---|---|---|
| B   0. | _____ | C   0. | _____ |
| 1. | _____ | 1. | _____ |
| 2. | _____ | 2. | _____ |
| 3. | _____ | 3. | _____ |
| 4. | _____ | 4. | _____ |
| 5. | _____ | 5. | _____ |
| 6. | _____ | 6. | _____ |
| 7. | _____ | 7. | _____ |
| 8. | _____ | 8. | _____ |
| 9. | _____ | 9. | _____ |
| 10. | _____ | 10. | _____ |
| 11. | _____ | 11. | _____ |
| 12. | _____ | 12. | _____ |
| 13. | _____ | 13. | _____ |

14. _____     14. _____

15. _____     15. _____

14. _____    14. _____

15. _____    15. _____

Name _____  Instructor _____

Partner _____

## REPORT SHEET

1. Distilled Water Samples:
   If the pH of your distilled water was not exactly 7, suggest an explanation for the value(s) you observed.

   _____

   _____

   ### Sample A

   Amount of added $0.1 M$ HNO$_3$ _____ moles

   Volume of sample _____

   Expected [H$^+$(aq)] _____

   Expected pH _____

   Observed pH _____

   ### Sample B

   Amount of added $0.1 M$ NaOH _____ moles

   Volume of sample _____

   Expected [H$^+$(aq)] _____

   Expected pH _____

   Expected pH _____

   Observed pH _____

   Suggest an explanation for any discrepancies between your observed and expected pH values.

2. Buffer Samples (check one):

   (Acetic acid–acetate, $K_a = 2.0 \times 10^{-5}$ _____)

   (Ammonia–ammonium, $K_b = 1.8 \times 10^{-5}$ _____)

   ### Sample A:

   | Volume H$_2$O added | Expected [H$^+$] | Expected pH | Observed pH |
   |---|---|---|---|
   | 0 mL | _____ | _____ | _____ |
   | 5 mL | _____ | _____ | _____ |
   | 10 mL | _____ | _____ | _____ |

15 mL          _____   _____   _____

20 mL          _____   _____   _____

Calculate the expected [H⁺(aq)] from $K_a$ or $K_b$ for one of the buffer components. Suggest an explanation for any observed discrepancies between observed and expected pH values.

### Sample B:

On a sheet of graph paper plot in black your values of observed pH against added 0.10$M$ HNO$_3$. From $K_a$ or $K_b$ and the Henderson–Hasselbach equation calculate the expected [H⁺(aq)] and pH for the addition of 1, 5, 7, 10 and 15 mL of added acid. Insert these calculated pH points in red on the same graph and suggest an explanation for any discrepancies between the two curves.

| Volume added acid | Expected [H⁺] | Expected pH | Observed pH |
|---|---|---|---|
| 1 mL | _____ | _____ | _____ |
| 5 mL | _____ | _____ | _____ |
| 7 mL | _____ | _____ | _____ |
| 10 mL | _____ | _____ | _____ |
| 15 mL | _____ | _____ | _____ |

Calculate the expected [H⁺(aq)] from $K_a$ or $K_b$ for one of the buffer components. Suggest an explanation for any observed discrepancies between observed and expected pH values.

### Sample C

On a sheet of graph paper plot in black your values of observed pH against added 0.10$M$ NaOH. From $K_a$ or $K_b$ and the Henderson–Hasselbach equation calculate the exptected [H⁺(aq)] and pH for the addition of 1, 5, 7, 10 and 15 mL of added base. Insert these calculated pH points in red on the same graph and suggest an explanation for any discrepancies between the two curves.

| Volume added base | Expected [H⁺] | Expected pH | Observed pH |
|---|---|---|---|
| 1 mL | _____ | _____ | _____ |
| 5 mL | _____ | _____ | _____ |
| 7 mL | _____ | _____ | _____ |
| 10 mL | _____ | _____ | _____ |
| 15 mL | _____ | _____ | _____ |

Calculate the expected [H⁺(aq)] from $K_a$ or $K_b$ for one of the buffer components. Suggest an explanation for any observed discrepencies between observed and expected pH values.

## WORKSHEET

Carry out all the calculations for Experiment 14(II) on this sheet, attach it to your report sheet, and submit the two together. Do your calculations directly on this sheet. Do not do them on scratch paper and then copy them on to this sheet. Make no erasures and delete incorrect entries with a single line so as to leave the original data legible. Use a pen.

# EXPERIMENT 15

# Introduction to Volumetric Analysis: Acid-Base Titrations in Aqueous Solution

### AIMS

To introduce the techniques and principles of volumetric analysis as applied to aqueous acid-base neutralization reactions. A stock solution of sodium hydroxide is standardized against potassium hydrogen phthalate, a primary standard. The base solution is then used to determine the unknown equivalent weight of an acid by titration.

### EQUIPMENT

From stockroom:
- 25-mL pipet
- 50-mL buret
- 250-mL volumetric flask

From locker:
- 500-mL glass bottle for acid solutions
- 500-mL plastic bottle for NaOH solution
- 50- and 100-mL beakers
- Erlenmeyer flasks (200-, 250- or 300-mL)
- Glass filter funnel
- Stirring rods

Laboratory:
- Buret stand and clamp
- Ring stand and clamp or tripod stand
- Burner and wire gauze

## REAGENTS

Potassium hydrogen phthalate, $KHC_8H_4O_4$
Stock sodium hydroxide solution
Assigned unknown solid acids

## BACKGROUND

### Associated Readings

The following sections of *Chemical Principles* should be read in conjunction with this Experiment 15.

**Chapter 15** Acids and Bases
   **15.1** Definition of Acids and Bases
   **15.2** Strengths of Acids and Bases
   **15.4** Acid-Base Reactions in Aqueous Solution
   **15.6** Neutralization and Titrations

Earlier in the course, in Experiment 3, we determined the empirical formula of a metal sulfide. The procedure was an example of **gravimetric analysis**—an analysis based on the masses of reactants and products in a reaction known to go effectively to completion. Gravimetric methods are widely used in chemical analysis and are intrinsically capable of yielding very accurate results. But, they are usually time consuming, always require a high degree of experimental skill, and are difficult to automate for routine use.

Where a rapid and experimentally straightforward method of analysis is needed that still yields accurate results, the chemist turns to **volumetric methods.** These methods make use of reactions that go rapidly to completion in solution. Where the concentration of a reagent solution is known, and where the stoichiometry of its reaction with a sample of a species of unknown concentration is known, the concentration of the sample is readily determined by establishing the volume required to exactly and completely react with a known volume of the reagent solution.

Reactions between acids and bases in aqueous solution, known as **neutralization reactions**, are important in volumetric analysis. We shall consider neutralization reactions as involving only the strongest acid and the strongest base in the system. (See Section 15.6 of *Chemical Principles*.) Three of the four classes of such reactions are of practical interest.

### 1. Strong Acid–Strong Base

Here the reaction of interest is:

$$H^+(aq) + OH^-(aq) \longrightarrow H_2O$$

for which $K = 1/K_w = 1.0 \times 10^{14}$ at room temperature, $K_w$ being the ion product for water.

### 2. Weak Acid–Strong Base

Here the reaction is:

$$HA(aq) + OH^-(aq) \longrightarrow A^-(aq) + H_2O$$

for which $K = K_a/K_w$, where $K_a$ is the dissociation constant for the acid.

### 3. Strong Acid–Weak Base

Here the reaction is:

$$H^+(aq) + B \longrightarrow HB^+(aq)$$

for which $K = K_b/K_w$, where $K_b$ is the base dissociation constant.

All of these reactions are very fast and the equilibrium constants normally sufficiently large to ensure that the reactions go effectively to completion. The fourth class, reactions between a weak acid and a weak base, for which $K = K_a K_b / K_w$ is often quite small, is not used in volumetric work.

The process by which one reactant is added to the other in a controlled fashion is known as a **titration.** The point at which a sufficient amount of the unknown solution has been added to react exactly with all of the known reagent (or vice versa) is known as the **equivalence point** of the reaction—the amounts of the two reactants are then stoichiometrically equivalent. The point at which this condition is experimentally determined to have been reached is known as the **end point** of the reaction. Ideally, the difference between these two points should be vanishingly small. In acid-base titrations, the attainment of the equivalence point is generally accompanied by a dramatic change in pH upon the addition of only one drop of excess reagent, and the end point is conveniently detected by indicators which reflect this change. The difference between the equivalence point and the end point, **the titration error,** is normally not significant for acid-base titrations in the three classes discussed above. Section 15.6 of *Chemical Principles* contains examples of how to calculate the pH at the equivalence point of an acid-base titration and you are asked to carry out a typical calculation of this kind as part of the prelaboratory preparation.

If we are dealing with acids which contain only one donatable proton per molecule and bases that can accept only one proton per molecule, the relationship between volumes ($V$) and concentrations ($M$) is

$$V_1 M_1 = V_2 M_2$$

Both volumes and one concentration are known so that the unknown concentration is determined and hence the unknown molecular weight.

However, many acids contain more than one available proton per molecule and some bases can accept more than one proton per molecule. When only one of the protons is titrated, what is derived from a titration is not the molecular weight of the unknown but its **equivalent weight.** The equivalent weight of an acid is its molecular weight divided by the total number of donatable protons and the equivalent weight of a base is its molecular weight divided by the total number of protons its molecule can accept.

$$\text{equivalent weight} = \frac{\text{molecular weight}}{\text{transferrable } H^+ \text{ per molecule}}$$

A more useful unit of concentration for acid-base titration work is **normality.** A normal solution (1N) contains the equivalent weight of the substance in one liter of solution. The relationship

$$V_1 N_1 = V_2 N_2$$

can always be used to find the equivalent weight of an unknown acid or base. To find the molecular weight from the equivalent weight we need to know the number of transferrable protons associated with the acid or the conjugate base.

## STANDARD SOLUTIONS

A **standard solution** is a solution of accurately known concentration. It might seem that the preparation of a standard solution would present few difficulties. If a pure stable material is available, simply weighing an accurately known mass of the substance and dissolving it to make a solution of accurately known volume is enough.

However, there are surprisingly few substances sufficiently pure and sufficiently stable, and of determinate composition, which can be used in this way as **primary standards.** For example, to illustrate the difficulties, sodium hydroxide and other caustic alkalis are hygroscopic at room temperature, i.e., they absorb water readily, and so cannot be weighed to any accuracy. Solutions of these bases must therefore be standardized against solutions of primary standards before their concentrations can be established. Because solutions of these alkalis attack glass, they must also be stored in plastic containers.

A primary standard widely used in acid-base work is solid potassium hydrogen phthalate, $KHC_3H_4O_4$, an acid with one available proton per molecule. This substance may be prepared to 99.97% purity, may be dried to constant weight, does not reabsorb water readily, is stable to oxidation, and is conveniently handled. The comparatively high molecular weight of the salt reduces the significance of small uncertainties in weighing, so that it may be used in comparatively small amounts.

In this experiment a standard solution of the phthalate (abbreviated KHP) is prepared and used to standardize (or accurately determine) the concentration of an approximately $0.1M$ solution of NaOH. The NaOH solution is then used to find the equivalent weight of an assigned acid sample by titration.

## PROCEDURE

Before reading further, turn to the Appendix and read the material on the design, use, and care of volumetric glassware. To obtain accurate results in volumetric analyses, you must always work with scrupulously clean equipment and exercise care and skill in the various manipulations called for. As in learning any new skill you may find your first few attempts awkward and clumsy, but with practice and experience you will soon be able to take pride in your technique and experience the confidence that comes from doing something really well.

### SAFETY NOTES
Observe proper precautions in handling acids and bases. Recall the proper procedures for filling pipets. Never fill a pipet by placing it in your mouth and sucking. Always use the bulb or tubing provided.

### 1. Preparation of the Standard KHP Solution

Set a clean, dry 100-mL beaker on the rough balance and add to it about 5 g of KHP. Take the beaker and its contents to the fine balance and weigh them to the nearest milligram.

Thoroughly clean the 250-mL volumetric flask following the instructions given in the Appendix. Place a clean, dry glass filter funnel in the mouth of the flask and carefully, without spills, transfer as much as possible of the KHP to the flask. Reweigh the empty beaker to the nearest milligram to establish the mass of KHP transferred. Use your wash bottle with distilled water to rinse any KHP lodged in the stem of the funnel into the flask. Continue rinsing the bowl and stem of the funnel, allowing the washings to pass into the flask, until the flask is about half full. Stopper the flask and swirl the contents thoroughly to dissolve the KHP completely. Make up the volume to within a few millimeters of the etched mark with distilled water and invert the flask several times to promote thorough mixing. Use your wash bottle in a drop-by-drop addition of distilled water to make up the solution exactly to the mark. Now with the flask firmly stoppered, invert the solution about two dozen times to ensure that it is homogeneous.

Because of the shape of the volumetric flask, it is difficult to obtain a completely homogeneous solution except by repeated inversions of the flask. At the discretion of your instructor you may adopt the following procedure to promote mixing. After having made up the solution to the mark, but before mixing it, transfer the contents of the volumetric flask to a clean, dry 500-mL glass bottle. Firmly stopper that bottle and gently invert the contents about a dozen times, with intermediate swirling. This procedure ensures better mixing but also risks incorporating $CO_2$ in the solution, which affects its acidity. If this procedure is adopted, it should be used each time an acid solution is prepared in an experiment so that any errors from $CO_2$ absorption will be about the same in the standardization and in any later titration and so cancel out.

## 2. Preparing the Buret

Clean the buret with a solution of detergent in cold distilled water and rinse it twice with distilled water alone. Allow the drainage to pass through the stopcock in each case so as to clean the tip as well as the main body of the buret.

Collect 500 mL of stock sodium hydroxide solution in a clean, dry plastic bottle. Transfer about 25 mL of the solution to a 50-mL beaker and use it to make three final rinses of the buret.

Now set up the buret in a buret stand and clamp. Load the buret with NaOH solution from the bottle, using a glass funnel to assist the transfer. Remove the funnel and open the stopcock so as to drain a few milliliters through the tip. Check that there is no air bubble in the tip and record the initial buret reading by sighting with the reading card prepared as part of the prelaboratory preparation. Read the buret to the nearest 0.03 mL.

## 3. Preparing the KHP Sample for Titration

Transfer about 20 mL of the KHP solution to a clean, dry 50-mL beaker and use it to make a final rinse of the previously cleaned pipet. (See Appendix.) Discard the rinsings.

Transfer about 40 mL of the KHP solution to a clean, dry 100-mL beaker. Pipet exactly 25 mL of that solution into a clean Erlenmeyer flask. Add two drops of phenolphthalein indicator to the contents of the flask and swirl to promote mixing. Use your wash bottle to rinse down the inside walls of the flask to get all the KHP solution in the body of the flask.

Titrate the solution in the flask against the NaOH, adding the first 15 mL or so of NaOH fairly quickly with swirling so as to promote mixing and reaction, and then more slowly as the end point approaches. As the end point nears, the pink color of the phenolphthalein that develops transiently as each drop of base enters the solution will become more persistent and finally remain as the first excess drop of base is added. There may be some fading of the color due to absorption into the solution of $CO_2$ from the air, but a color persisting for 15 seconds is a sufficient indication that the end point has been reached. Be sure to check the end point by rinsing down with distilled water any drops of unreacted solution that may be sticking to the inside walls of the flask.

Repeat the procedure with a second 25-mL portion of KHP solution. The volumes of NaOH used in each titration should agree to within $\pm 0.05$ mL but, as it is easy to overshoot the end point on the first titration, a third titration should be carried out if the first two results do not so agree. The later titrations may be carried out much more quickly as you now know the approximate location of the end point.

Once you have obtained two consecutive titrations agreeing to within $\pm 0.05$ mL, discard the KHP solution and wash and dry the glass bottle and its stopper. A final rinse of the bottle with acetone may be needed to get the bottle properly dry. It must be dry before it is used again for mixing purposes.

## 4. Determination of the Equivalent Weight of the Unknown Acid

Your instructor will assign you a sample of unknown solid acid. Follow the procedure used for preparing the KHP solution but weigh out about 1.5 g of the unknown acid (to the nearest milligram) to make up to 250 mL in the volumetric flask. Titrate 25-mL portions of this solution against NaOH. Consecutive readings should agree to ±0.05 mL.

## 5. Final Things

Discard any remaining KHP or unknown acid solutions into the sink, flushing them down with copious amounts of cold water. Clean and rinse with cold water all the volumetric glassware used, being sure to clean around the stopcock of the buret. Return the volumetric ware to the stockroom. Carefully stopper and label the plastic bottle containing the sodium hydroxide solution and store it in your locker or in the designated area. Be sure that the bottle carries your name if it is stored outside the locker. The NaOH solution will be used again in Experiment 16.

3. Salicylic acid, a monoprotic organic acid, has $K_a = 1.1 \times 10^{-3}$ at 25°C. What is the pH at the equivalence point when a 0.10$M$ solution of salicylic acid is titrated against 0.10$M$ NaOH?

___

___

___

___

4. Recalling Experiment 14.1, suggest a suitable indicator to detect the end point of the reaction in Item 3 above.

___

5. To help in reading the buret in this experiment construct a reading card, as described in the Appendix. Show the card to your instructor when you come to laboratory and before you use it.

## WORKSHEET

Carry out all the calculations for Experiment 15 on this sheet and attach it to your report sheet, before submitting and submit the two together. Do your calculations directly on this sheet. Do not do them on scratch paper and then copy them on to this sheet. Make no erasures and delete incorrect entries with a single line so as to leave the original data legible. Use a pen.

# EXPERIMENT 16

# Determination of Antacid Action

### AIM

To determine the antacid capacity of some commercial preparations.

### EQUIPMENT

| | |
|---|---|
| From stockroom: | 5- and 25-mL pipets |
| | 50-mL buret |
| | 250-mL volumetric flask |
| From locker: | 30-mL beaker |
| | Erlenmeyer flasks (200-, 250-, or 300-mL) |
| | Glass filter funnel |
| Laboratory: | Buret stand and clamp |
| | Tripod stand |
| | Burner |
| | Wire gauze |
| | Glass wool |

## REAGENTS

6$M$ hydrochloric acid
Various antacid tablets
Bromthymol blue indicator
Phenolphthalein indicator

## BACKGROUND

### Associated Readings

The following sections of *Chemical Principles* should be read in conjunction with Experiment 16.

**Chapter 15** Acids and Bases
        **15.1** Definitions of Acids and Bases
        **15.2** Strengths of Acids and Bases
        **15.4** Acid-Base Reactions in Aqueous Solutions
        **15.6** Neutralization and Titrations

Stomach ulcers and the general digestive disorders loosely grouped under the heading of hyperacidity are not new, although our scientific understanding of the mechanism of digestion and its defects is of fairly recent origin. A landmark episode in the development of this understanding was provided inadvertently by a Canadian fur trapper, Alexis St. Martin, who in 1822 had a hole blown in his side by a shotgun blast. He was treated by an army doctor, William Beaumont, and survived. The wound healed but left a hole about 2 cm in diameter leading through St. Martin's side directly into his stomach. Beaumont, working in very primitive conditions, and with the grudging cooperation of his patient, was able to use St. Martin as a human guinea pig and made a remarkable series of observations of stomach processes at work *in situ*.

Food, when chewed and swallowed, is directed by powerful muscular action down the esophagus into the stomach. So powerful is this muscular action that you can actually eat in an upside down position if you practice. The stomach is located not, as is widely supposed, behind the navel but under the rib cage on the left side—heart burn is really stomach burn. In the stomach food is "mashed, churned, pulverized, and generally battered beyond recognition," to quote our source. Besides this mechanical action the stomach acts chemically to digest its intake. More than a score of enzymes, chief among them being pepsin, acting in a solution of hydrochloric acid, rip apart the constituent macromolecules of the food breaking them down into ever smaller fragments and liberating a great deal of energy. An average daily intake of food produces close to 12,000 kJ of energy.

The hydrochloric acid concentration in the stomach is high—about 0.15$M$; the solution is strong enough to dissolve a cotton rag—and the stomach is protected against the acid by a sticky mucous lining which is constantly being destroyed and replaced.

From the stomach the partially digested food passes to the intestinal tract through the pylorus, a muscle that controls the rate at which food is transferred. People who feel perpetually hungry often have a defective pylorus which allows food to pass too quickly from the stomach. In the small intestine the acid solution generated in the stomach is neutralized by alkaline enzyme secretions produced in the pancreas and liver. This process occurs in the duodenum, and it is in the intestine, with the acidity neutralized, that the process of enzymatic digestion begins in earnest. Here, proteins are broken down into their constituent amino acids, carbohydrates to their constituent sugars, and fats to fatty acids and glycerol.

Metabolically useful molecules are scavenged by the villi, microscopically small whiskers lining the inner intestinal walls, and are fed into the various metabolic chains that sustain life. The waste molecules are further broken down and are pushed by peristalsis into the colon, eventually to be passed out of the body through the rectum.

The digestive process is under the control of the autonomous nervous system, and save for the occasional fakir, we are unable consciously to affect its functioning. It is, however, subject to disturbance if our emotional state is in any way out of equilibrium. Anger, fear, tension, depression, and tranquillity are all reflected in the digestion. Beaumont directly observed the changes in stomach action attendant upon states of anger and excitement in St. Martin and, among other conclusions, noted that "the constant and persistent use of ardent spirits leads to the complete destruction of the stomach." Each of you is probably familiar with the dry mouth of fear, the leaden stomach of depression, the cramps of tension, the glow that follows a good meal, and so on. The stomach and the digestion are the barometer of the emotions, the true mirror of the soul.

A further advance in our understanding followed another accident, this time to a boy named Tom who, in Boston in 1895, took a large gulp of scalding hot clam chowder in the mistaken belief that it was cold beer. Tom's esophagus was permanently blocked by this mischance and a gastrostomy was performed whereby a hole was opened directly into his stomach, through which a tube could be inserted for the direct passage of food. For over fifty years Tom fed himself by chewing his food and spitting it into a funnel attached to his tube. Doctors noted that when Tom was angry his gastric juices multiplied, whereas when he was depressed his gastric activity slowed. Digestive hyperactivity leads to excess production of stomach acid which, while it normally does no real harm so long as it remains in the stomach and the condition is not chronic, can cause distress if it escapes from the stomach. It may escape in either of two directions: back into the esophagus with a gas bubble, leading to heartburn; or forward into the unprotected intestine where it may lead to ulcers by eating its way through the lining of the duodenum. Since, in cases of extreme emotion, more than twice the normal quantity of acid actually needed for digestion may be produced, the excess spends itself in eating away at the stomach lining or in other harmful ways.

Given that ulcers and other digestive upsets are primarily emotional in origin, the acid acts merely as an agent, and given the many sources of stress in today's world, it is scarcely surprising that multitudes of people suffer, or feel that they suffer, from these ailments. Equally, with the habit of treating symptoms rather than prime causes firmly established in our society, it is not surprising that millions of dollars worth of antacid remedies are sold each week. Although it is doubtful that these preparations offer any long-term benefits to the individual, there is no doubt that they do react with acids and that they do partially reduce excess stomach acid when taken in the recommended doses. It is to be recognized, of course, that so long as the emotional trigger remains active the stomach will continue to respond by producing additional acid, so that even the short term benefits are as likely to be psychological as physiological in origin. No matter, the end effect is the same.

The majority of patent medicine antacids are carbonates or bicarbonates of calcium or sodium, with the occasional preparation containing hydroxides of aluminum or magnesium. Many are mixtures of several of these ingredients and most contain sweeteners and glues or other binders.

The action of calcium carbonate with hydrochloric acid proceeds according to the equation:

$$CaCO_3(s) + 2H^+(aq) \longrightarrow Ca^{2+}(aq) + CO_2(g) + H_2O$$

The carbon dioxide produced may ease discomfort by provoking belching, thus reducing gas pressure in the stomach, but the liberated gas may carry acid into the esophagus.

The hydroxide-based preparations react to form only water and so do not lead to that gassy feeling.

$$Al(OH)_3(s) + 3H^+(aq) \longrightarrow Al^{3+}(aq) + 3H_2O$$

In this experiment we will determine the amount of HCl neutralized by a single tablet dose of some common commercial preparations and will refer their antacid capacity to a common base in terms of the amount of calcium carbonate corresponding to their active components. Such information, combined with recommended dosage levels and price, allows an informed consumer judgment as to which preparations are to be preferred.

Solution of the tablets in stomach acid is slow. We will use the much stronger $6M$ HCl to dissolve the tablets, but since we do not wish to carry out titrations with acid of this strength, after the tablet has dissolved we will dilute the residual acid by a known factor and establish its concentration by back titration.

## PROCEDURE

### SAFETY NOTES

Treat the $6M$ HCl with care and respect. Follow the prescribed rules for pipetting this acid. Under no circumstances should you attempt to pipet this acid by placing the pipet directly in your mouth.

The antacid tablets are provided for chemical use only. They are not to be ingested.

### 1. Determination of the Exact Concentration of the $6M$ HCl

The $6M$ HCl supplied is only approximately that strength. We shall dilute it by a factor of 50 and establish the concentration of the diluted acid by titration against standard NaOH.

Use a small amount of the $6M$ acid in a clean 30-mL beaker to rinse the 5-mL pipet. Discard the excess acid, washing it down the sink with copious amounts of cold water. From another sample of the acid in a clean, dry 30-mL beaker, pipet exactly 5 mL of the acid into a clean 250-mL volumetric flask containing about 50 mL of distilled water. Mix the contents of the flask well and make up to the mark with distilled water. Complete the mixing in the usual way.

Titrate 25-mL aliquots of the diluted acid against your standard NaOH solution using 2 drops of phenolphthalein as indicator. The titration volumes of NaOH should agree to within 0.10 mL.

### 2. Preparation of the Sample

Using the rough balance, establish the mass of the antacid tablet of your choice to the nearest 10 mg. Transfer the tablet to a clean Erlenmeyer flask and add to the flask about 5 mL of distilled water. Evolution of gas bubbles at this stage indicates that the tablet contains bicarbonate. When any evolution of gas has ceased, pipet exactly 5 mL of the $6M$ HCl into the flask containing the tablet and water. Note on your data sheet whether or not a gas is evolved. When the tablet has dissolved, carefully add 50 mL of distilled water. Although the tablet will have neutralized most of the acid, you are still adding water to acid and should carry out this step with care.

Add one drop of bromthymol blue indicator to the solution to make sure that an excess of acid is present. If the indicator shows the solution to be alkaline, pipet another 5 mL portion of $6M$ acid to the flask and swirl the contents.

Boil the solution gently for about five minutes to drive off any dissolved carbon dioxide and to complete the reaction. Allow the solution to cool to room temperature.

If the cool solution is clear, transfer it to a clean 250-mL volumetric flask. Wash the insides of the Erlenmeyer flask thoroughly with distilled water, transferring the washings to the volumetric flask so as to ensure that all the acid is transferred. Make up to the mark with distilled water.

If the solution is cloudy, owing to the presence of undissolved binder or filler, make the transfer to the volumetric flask through a glass wool plug in the filter funnel. Some very finely divided material will pass through but should not affect the subsequent titrations.

## 3. Titration of the Sample with NaOH

Titrate 25-mL aliquots of the diluted sample preparation against your standard NaOH. Readings should agree to within 0.10 mL. Choose an indicator whose color change is not masked by any color in the sample solution.

Repeat Steps 2 and 3 with as many different preparations as time permits. Space is provided on the data sheet for three samples.

Name _____  Instructor _____

## PRELABORATORY PREPARATION SHEET

1. Do some independent reading to supplement the background material and find out the average volume and concentration of stomach HCl produced to digest an average meal.

   _____

   [Source] _____

2. From this same reading, identify some other enzymes besides pepsin involved in the digestive process.

   _____

   _____

   [Source] _____

3. Why do you suppose that NaOH, KOH, and $Ca(OH)_2$ are not used as antacids?

   _____

4. A 5.00-mL volume of approximately $6M$ HCl is diluted to 250 mL with distilled water in a volumetric flask. A 25.00-mL aliquot of the diluted solution requires exactly 24.00 mL of $0.150M$ NaOH for neutralization. What is the actual concentration of the original HCl? Show your calculation.

   _____

   _____

   _____

   _____

   _____

347

Name _____  Instructor _____

## DATA SHEET

1. General observations:

| | Name | Color | Gas evolved with $H_2O$ | Gas evolved with acid |
|---|---|---|---|---|
| Sample 1 | _____ | _____ | _____ | _____ |
| Sample 2 | _____ | _____ | _____ | _____ |
| Sample 3 | _____ | _____ | _____ | _____ |

2. Titration of 6$M$ HCl (25 mL, 5/250 dilution):

   Final buret reading        _____   _____   _____

   Initial buret reading      _____   _____   _____

   Volume NaOH delivered      _____   _____   _____

   Mean volume NaOH           _____ ± _____

3. Sample masses:

   |   | 1 | 2 | 3 |
   |---|---|---|---|
   |   | _____ | _____ | _____ |

4. Sample titrations (25 mL, 5/250 dilution):

### Sample 1

   Final buret reading        _____   _____   _____

   Initial buret reading      _____   _____   _____

   Volume NaOH delivered      _____   _____   _____

   Mean volume NaOH           _____ ± _____

### Sample 2

   Final buret reading        _____   _____   _____

   Initial buret reading      _____   _____   _____

   Volume NaOH delivered      _____   _____   _____

   Mean volume NaOH           _____ ± _____

### Sample 3

   Final buret reading        _____   _____   _____

   Initial buret reading      _____   _____   _____

   Volume NaOH delivered      _____   _____   _____

   Mean volume NaOH           _____ ± _____

Name _____ Instructor _____

# DATA SHEET

1. General observations:

|  | Name | Color | Gas evolved with $H_2O$ | Gas evolved with acid |
|---|---|---|---|---|
| Sample 1 | _____ | _____ | _____ | _____ |
| Sample 2 | _____ | _____ | _____ | _____ |
| Sample 3 | _____ | _____ | _____ | _____ |

2. Titration of $6M$ HCl (25 mL, 5/250 dilution):

| | | | |
|---|---|---|---|
| Final buret reading | _____ | _____ | _____ |
| Initial buret reading | _____ | _____ | _____ |
| Volume NaOH delivered | _____ | _____ | _____ |
| Mean volume NaOH | _____ ± _____ | | |

3. Sample masses:

| | 1 | 2 | 3 |
|---|---|---|---|
| | _____ | _____ | _____ |

4. Sample titrations (25 mL, 5/250 dilution):

## Sample 1

| | 1 | 2 | 3 |
|---|---|---|---|
| Final buret reading | _____ | _____ | _____ |
| Initial buret reading | _____ | _____ | _____ |
| Volume NaOH delivered | _____ | _____ | _____ |
| Mean volume NaOH | _____ ± _____ | | |

## Sample 2

| | 1 | 2 | 3 |
|---|---|---|---|
| Final buret reading | _____ | _____ | _____ |
| Initial buret reading | _____ | _____ | _____ |
| Volume NaOH delivered | _____ | _____ | _____ |
| Mean volume NaOH | _____ ± _____ | | |

## Sample 3

| | 1 | 2 | 3 |
|---|---|---|---|
| Final buret reading | _____ | _____ | _____ |
| Initial buret reading | _____ | _____ | _____ |
| Volume NaOH delivered | _____ | _____ | _____ |
| Mean volume NaOH | _____ ± _____ | | |

**5.** Other Observations:

|  | 1 | 2 | 3 |
|---|---|---|---|
| Cost per tablet | _____ | _____ | _____ |
| Dose suggested | _____ | _____ | _____ |
| Solid residue | _____ | _____ | _____ |

5. Other Observations:

|  | 1 | 2 | 3 |
|---|---|---|---|
| Cost per tablet | | | |
| Dose suggested | | | |
| Solid residue | | | |

Name _____ Instructor _____

# REPORT SHEET

1. Enter the mean vol. of NaOH required for titration of 25 mL diluted HCl. _____

2. Enter the molarity of your NaOH solution. _____

3. Calculate the molarity of the diluted HCl. _____
4. Apply the dilution factor to find the molarity of the original 6M HCl.
   _____

5. For each sample give the mean volume of NaOH used in the titration, the molarity of the NaOH, and the calculated molarity of the diluted acid titrated. Apply the dilution factor (250/5) to find the molarity of the *undiluted* residual HCl after reaction with the sample.

   | Sample | 1 | 2 | 3 |
   |---|---|---|---|
   | Volume of NaOH | | | |
   | Molarity | | | |
   | Molarity of diluted HCl | | | |
   | Molarity of undiluted HCl | | | |

6. From the molarity of the original HCl, calculate the moles of HCl present in 5 mL of the acid.
   _____

7. From the molarity of the undiluted HCl after reaction with the sample, calculate the moles of HCl remaining.

   Sample

   1 _____

   2 _____

   3 _____

8. Give the moles of acid consumed by each sample.

   Sample

   1 _____

   2 _____

   3 _____

9. Write a balanced equation for the reaction of CaCO$_3$ with HCl.
   _____

10. On the assumption that the active ingredient of each tablet is $CaCO_3$, calculate the mass in mg of $CaCO_3$ in each tablet.

   **Sample**

   1 _____

   2 _____

   3 _____

11. Calculate the cost per 100 mg of $CaCO_3$.

   **Sample**

   1 _____

   2 _____

   3 _____

12. In the prelaboratory preparation you were asked to find how much stomach acid was produced to digest a typical meal. Suppose that a 50% excess of stomach acid is produced in response to the realization that you have just called your aunt's favorite Pekinese a mongrel, in her hearing, and that she has announced she is cutting you out of her will. How many tablets of Sample 1 would you require to take in order to neutralize this *excess* completely? How does this number compare to the recommended dosage?

   _____

   _____

   _____

   _____

   _____

## WORKSHEET

Carry out all the calculations for Experiment 16 on this sheet, attach it to your report sheet, and submit the two together. Do your calculations directly on this sheet. Do not do them on scratch paper and then copy them on to this sheet. Make no erasures and delete incorrect entries with a single line so as to leave the original data legible. Use a pen.

# EXPERIMENT 17

# Oxidation-Reduction Reactions of Vanadium in Acid Solution

### AIM

To establish the effects of two reducing agents, $SO_2$ and Zn, on an acidic solution of dioxovanadium ion, $VO_2^+$.

### EQUIPMENT

From stockroom: 50-mL buret
From locker: Erlenmeyer flasks (200-, 250-, or 300-mL)
25-mL graduated cylinder
30-mL beaker
Glass filter funnel
Thermometer
Laboratory: Buret stand and clamp
Tripod or ring stand
Burner and wire gauze
Filter paper or glass wool

## REAGENTS

Stock solution of $VO_2^+$  
2$M$ and 6$M$ sulfuric acid  
Sodium sulfite, $Na_2SO_3$  
Stock solution of $KMnO_4$  
1$M$ hydrochloric acid  
Ferric ammonium sulfate,  
  $NH_4Fe(SO_4)_2 \cdot 12H_2O$

## BACKGROUND

### Associated Readings

The following sections of *Chemical Principles* should be read in conjunction with Experiment 17.

**Chapter 16** Oxidation-Reduction
  **16.1** Oxidation-Reduction Reactions
  **16.2** Oxidation-Reduction Processes in Aqueous Solution
  **16.6** Electrode Potentials

In acidic aqueous solutions vanadium can exist in four different oxidation states. In the two highest of these, vanadium(V) and vanadium(IV), it is present as the oxygenated species $VO_2^+$—the dioxovanadium ion (in its hydrated form sometimes written $V(OH)_4^+$), and $VO^{2+}$—the vanadyl cation. In the two lower states, vanadium(III) and vanadium(II) it occurs as the hydrated $V^{3+}$ and $V^{2+}$ cations. Each of these four species has a characteristic color in solution, and the oxidation state of vanadium in a given compound may often be deduced directly on that basis alone.

The dioxovanadium ion, $VO_2^+$, may be reduced chemically to lower oxidation states of vanadium. The standard electrode potentials associated with the reduction half-reactions of vanadium species are:

| Half-reaction | $\varepsilon°$ (V) |
|---|---|
| $VO_2^+(aq) + 2H^+(aq) + e^- \longrightarrow VO^{2+}(aq) + H_2O$ | 1.00 |
| $VO^{2+}(aq) + 2H^+(aq) + e^- \longrightarrow V^{3+}(aq) + H_2O$ | 0.36 |
| $V^{3+}(aq) + e^- \longrightarrow V^{2+}(aq)$ | $-0.26$ |

The oxidation state to which $VO_2^+$ is reduced in a redox reaction depends on the strength of the reducing agent used, a stronger reducing agent being needed to bring about reduction to V(II) and V(III) than to V(IV).

In this experiment, equal volumes of a solution containing $VO_2^+$ ion will be separately reduced with zinc metal and with sulfur dioxide, and the resulting oxidation states of vanadium determined.

The oxidation half-reactions for zinc and for sulfur dioxide, and their associated standard oxidation potentials, are:

| | $\varepsilon°$ (V) |
|---|---|
| $Zn \longrightarrow Zn^{2+}(aq) + 2e^-$ | 0.76 |
| $SO_2(aq) + 2H_2O \longrightarrow SO_4^{2-}(aq) + 4H^+(aq) + 2e^-$ | $-0.17$ |

Because we are using solutions in which the concentrations of the various species are far from 1$M$, and because kinetic factors often outweigh thermodynamic ones, we cannot use the standard potentials to predict unequivocally which reactions will take place when either $SO_2$ or Zn is used to reduce the $VO_2^+$ solution. Note, however, that the oxidation half-potentials indicate that zinc is a stronger reducing agent than sulfur dioxide.

The state to which vanadium is reduced in each case is determined by comparison of the volumes of standard permanganate solution needed to reoxidize the reduced vanadium species produced when equal volumes of the same $VO_2^+$ solution are treated with zinc and with sulfur dioxide.

Permanganate anion, $MnO_4^-$, is a very strong oxidizing agent. In acid solution it will quantitatively convert any reduced vanadium species to the fully oxidized vanadium(V) state as $VO_2^+$. In the process permanganate is reduced to yield manganese(II) as the manganous ion, $Mn^{2+}$. The half-reaction is:

$$MnO_4^-(aq) + 8H^+(aq) + 5e^- \longrightarrow Mn^{2+}(aq) + 4H_2O, \quad \varepsilon° = 1.51 \text{ V} \quad (1)$$

The amount of permanganate consumed in a redox reaction is proportional to the moles of electrons transferred in the reaction. For stoichiometrically equal amounts of vanadium present in different oxidation states, the amount of permanganate needed to bring about complete reoxidation to vanadium(V) will be directly proportional to the moles of electrons needed for reoxidation. For the same permanganate solution, the volumes of permanganate needed will thus stand in a simple numerical ratio from which the oxidation states involved may be deduced. Equal amounts of vanadium(IV) and vanadium(III), for example, would require twice the volume of permanganate to reoxidize the latter as the former. Once the vanadium oxidation states are known, balanced equations may be written for the overall reduction process and for the reoxidations, and the standard cell potentials calculated.

The simple method outlined is complicated by the fact that the low oxidation state to which zinc reduces $VO_2^+$ is unstable in air, with partial reoxidation of the reduced vanadium taking place by reaction with dissolved oxygen. One way of avoiding this problem would be to conduct the experiment under an inert atmosphere such as nitrogen. Such a technique is used routinely in many chemical laboratories where air-sensitive materials are handled, but it lies outside the scope of this course. Instead, we avoid the difficulty by deliberately oxidizing the vanadium species produced on zinc reduction to a higher air-stable form, using an oxidizing agent which is itself reduced to a more stable form than the vanadium species. This is done quantitatively by addition of excess ferric ion, $Fe^{3+}$. The ferric ion is reduced to the ferrous state as indicated:

$$nFe^{3+}(aq) + mV^{x+}(aq) \longrightarrow nFe^{2+}(aq) + mV^{y+}(aq) \quad (2)$$

and the permanganate is used to reoxidize both $V^{y+}$ to $VO_2^+$ and $Fe^{2+}$ to $Fe^{3+}$. Because the number of electrons transferred in the reoxidation of $Fe^{2+}$ to $Fe^{3+}$ is exactly the same as that transferred in the oxidation of $V^{x+}$ to $V^{y+}$, the permanganate titration effectively establishes the number of electrons needed to carry out the oxidation $V^{x+}$ to $VO_2^+$. The detailed working out of this argument is required in the prelaboratory preparation sheet.

## PROCEDURE

### SAFETY NOTES
Sulfur dioxide is a highly irritant gas. The $SO_2$ needed for the reaction is generated directly in the reaction mixture by addition of sodium sulfite to the acidified $VO_2^+$ solution. Excess $SO_2$ is boiled off, so the reaction with sulfite should be carried out under a hood.

### Sulfur Dioxide Reduction

1. Measure 25 mL of the stock $VO_2^+$ solution into your graduated cylinder and transfer the solution to a clean, dry 250-mL Erlenmeyer flask. Without rinsing the graduated cylinder, use it to collect 10 mL of 6M sulfuric acid and transfer the acid to the same flask. Swirl the solution to ensure mixing.

2. Set up a buret and stand and ready the buret with the stock permanganate solution.
3. On the rough balance weigh out about 0.20 g of sodium sulfite into a clean, dry 30-mL beaker. At the hood transfer the sulfite to the flask containing the acidified $VO_2^+$ solution and swirl the contents of the flask carefully until all of the solid has dissolved and the color of the solution has stabilized. On your data sheet note the color changes that take place during the addition and mixing.

Add a boiling chip to the solution, set the flask on a tripod stand with wire gauze and, under the hood, gently boil the solution for about 10 minutes to drive off excess $SO_2$. Take care that no solution is lost by spurting and that the volume of the solution is not too much reduced by evaporation. Allow the solution to cool to about 80°C, remove the flask from the hood (*no tongs, please*) and quickly titrate the contents against permanganate. Note any color changes during the titration and record the volume of permanganate needed.

## Zinc Reduction

1. Weigh about 7 g of granulated zinc into a clean Erlenmeyer flask.
2. Prepare 100 mL of a 0.066$M$ solution of ferric ammonium sulfate. (You calculate the mass of solid needed for this step as part of the prelaboratory preparation.) Pour about 60 mL of this solution into a separate Erlenmeyer flask. Note the color of the solution.
3. With the graduated cylinder, transfer 25 mL of the stock $VO_2^+$ solution to the flask containing the zinc granules and follow this with 5 mL of 2$M$ sulfuric acid. When any vigorous effervescence has ceased, stopper the flask and swirl the mixture until the solution is a clear violet color. Note the sequence of color changes that takes place during the reduction. The reaction takes about 10 minutes to go to completion and it is important to allow it to reach completion.
4. Top up the buret with stock permanganate solution.
5. With a ring stand set up over the flask containing the solution of ferric ammonium sulfate a glass filter funnel and filter paper. Wet the filter paper with distilled water, using your washbottle, and then rinse the filter paper with 5 mL of 2$M$ $H_2SO_4$.

    When the zinc reduction is complete, rapidly transfer the reduced solution to the flask containing the ferric ammonium sulfate, using the filter to trap any unreacted zinc. Again note the color changes taking place in the flask. Thoroughly rinse the flask that contained the reduced solution with three 25-mL lots of 2$M$ $H_2SO_4$, adding the washings through the filter to the other flask.
6. Warm the solution to be titrated to about 80°C and titrate it against the stock permanganate. By following the color changes in the flask as the titration proceeds (they should be the reverse of those noted during the reduction), you can estimate when you approach the end point. Record the color changes and the volume of permanganate needed for titration to $VO_2^+$.

Before leaving the laboratory at the end of this period, fill your glass bottle to the halfway mark with the stock permanganate solution, label the bottle with your name, and store it in your locker or in the designated area. This will give the permanganate time to stabilize in effective concentration before you standardize it in Experiment 19.

Name _____  Instructor _____

## PRELABORATORY PREPARATION SHEET

1. Calculate the mass of ferric ammonium sulfate, $NH_4Fe(SO_4)_2 \cdot 12H_2O$ needed to make 100 mL of a $0.066M$ solution.

   _____

   _____

2. Why is it important that no metallic zinc be transferred to the solution to be titrated in Step 6 of the zinc reduction procedure?

   _____

3. Why is it necessary to continue to boil the acidified sulfite solution after the reduction is apparently complete?

   _____

4. No indicators are specified for the permanganate titrations. How do you suppose you will recognize the end points?

   _____

   _____

5. Consider Equations (1) and (2) in the background material. Write equations using letter coefficients and excluding protons for the titration of $Fe^{2+}$ and $V^{y+}$ separately with permanganate. Show how, by combination of these equations with Equation (2), the overall equation for the reoxidation of $V^{x+}$ with $MnO_4^-$ may be derived.

   _____

   _____

   _____

   _____

   _____

Name _____  Instructor _____

## DATA SHEET

|  | SO$_2$ reduction | Zn reduction |
|---|---|---|
| Volume VO$_2^+$ solution | _____ | _____ |
| Initial color of solution | _____ | _____ |
| Color changes on reduction | _____ | _____ |
|  | _____ | _____ |
|  | _____ | _____ |
|  | _____ | _____ |
| Color changes on oxidation | _____ | _____ |
|  | _____ | _____ |
|  | _____ | _____ |
|  | _____ | _____ |
| Volume MnO$_4^-$ needed in titration | _____ | _____ |
| Simplified ratio of volumes | _____ | _____ |

Name _____  Instructor _____

## DATA SHEET

|  | $SO_2$ reduction | Zn reduction |
|---|---|---|
| Volume $VO_2^+$ solution | _____ | _____ |
| Initial color of solution | _____ | _____ |
| Color changes on reduction | _____ | _____ |
|  | _____ | _____ |
|  | _____ | _____ |
|  | _____ | _____ |
| Color changes on oxidation | _____ | _____ |
|  | _____ | _____ |
|  | _____ | _____ |
|  | _____ | _____ |
| Volume $MnO_4^-$ needed in titration | _____ | _____ |
| Simplified ratio of volumes | _____ | _____ |

Name _____  Instructor _____

# REPORT SHEET

1. Report the simplified ratio of the volumes of permanganate needed in the titrations

   ($SO_2$:Zn) _____

2. Sulfur dioxide reduction:
   a. Give the formula of the vanadium species formed on reduction.

   _____

   b. Write a balanced net ionic equation for the reduction reaction.

   _____

   c. Write a balanced net ionic equation for the reoxidation of this reduced vanadium species by permanganate.

   _____

3. Zinc reduction:

   a. Give the formula of the vanadium species resulting from the reduction. _____
   b. Write a balanced net ionic equation for the reduction reaction.

   _____

   c. Write a balanced net ionic equation for the reoxidation of this reduced vanadium species by permanganate.

   _____

4. Color assignments:
   a. From your recorded observations on color changes, assign a color to each of the following vanadium species.

   $VO_2^+$ _____  $V^{3+}$ _____

   $VO^{2+}$ _____  $V^{2+}$ _____

   b. Account for any other colors observed.

   _____

   c. On the basis of these assignments, identify the unknown oxidation states of vanadium in the intermediate oxidation step $V^{x+} + Fe^{3+} \rightarrow V^{y+} + Fe^{2+}$.

   $x =$ _____ Ion _____  $y =$ _____ Ion _____

   d. Write a balanced net ionic equation for this intermediate step.

   _____

373

## WORKSHEET

Carry out all the calculations for Experiment 17 on this sheet, attach it to your report sheet, and submit the two together. Do your calculations directly on this sheet. Do not do them on scratch paper and then copy them on to this sheet. Make no erasures and delete incorrect entries with a single line so as to leave the original data legible. Use a pen.

# EXPERIMENT 18

# Synthesis of a Potassium Oxalatoferrate Complex

### AIM

To synthesize an iron(III)-containing oxalate complex, whose empirical formula will later be determined by permanganate titration.

### EQUIPMENT

| | |
|---|---|
| From stockroom: | 50-mL buret |
| | Buchner filter funnel and flask |
| From locker: | 30- and 250-mL beakers |
| | Glass filter funnel |
| | Thermometer |
| | Watchglass |
| Laboratory: | 50-mL graduated cylinder |
| | Filter paper |

## REAGENTS

Ferrous ammonium sulfate, $Fe(NH_4)_2(SO_4)_2 \cdot 6H_2O$
6% aqueous hydrogen peroxide, $H_2O_2$
1$M$ oxalic acid, $H_2C_2O_4$
2$M$ potassium oxalate, $K_2C_2O_4$
6$M$ sulfuric acid, $H_2SO_4$
Ethanol and acetone

## BACKGROUND

### Associated Readings

The following sections of *Chemical Principles* should be read in conjunction with Experiment 18.

**Chapter 18** Coordination Chemistry
        18.1 Structure of Coordination Compounds
        18.2 Isomerism in Coordination Compounds
        18.5 Complex Ions in Aqueous Equilibria

A **coordination complex** is a compound in which a metal ion forms coordinate bonds with one or more electron-pair donors. The donating groups are called **ligands** and the resulting complex may be charged or electrically neutral, depending on the number and charges of the ligands and the initial charge on the metal. As examples, the copper(II) ion, $Cu^{2+}$, forms a cationic ammine complex with four molecules of ammonia:

$$Cu^{2+}(aq) + 4NH_3(aq) \longrightarrow [Cu(NH_3)_4]^{2+}(aq)$$

an anionic complex with chloride ion:

$$Cu^{2+}(aq) + 4Cl^-(aq) \longrightarrow [CuCl_4]^{2-}(aq)$$

and a neutral complex with amino acids such as glycine:

$$Cu^{2+}(aq) + 2H_2N \cdot CH_2COOH(aq) \longrightarrow Cu(HN \cdot CH_2COO)_2(aq) + 2H^+(aq)$$

The **coordination number** of copper, the number of coordinate bonds formed by the copper ion, is four in each of these cases, though much weaker coordinate bonds from water molecules may increase that number to six in the first and third cases. In the ammine and chloride cases, each ligand forms one bond to the metal ion. In the case of the glycine complex, each ligand forms two bonds with the metal ion as illustrated by the structural formula:

# Synthesis of a Potassium Oxalatoferrate Complex

Complexes in which the ligand forms more than one coordinate bond with the metal are known as **chelates,** and the ligands are often referred to as **chelating agents.** A ligand which forms only one bond with the metal ion is known as a **monodentate** ligand. Ligands forming two, three, four, or six bonds with the same metal ion are known as **bi-, ter-, quadri-,** or **hexadentate** ligands, respectively.

In this experiment we form a complex between the oxalate anion, $C_2O_4^{2-}$, and $Fe^{3+}$. The resultant complex is anionic, and we isolate it as its hydrated potassium salt.

The salt has the formula $K_w[Fe_x(C_2O_4)_y] \cdot zH_2O$, where $w$, $x$, $y$, and $z$ are to be determined in a later experiment. From the formula for this salt it is evident that complexes containing more than one acceptor metal ion are possible. Such complexes are known as **bi-** or **trinuclear** complexes, etc., depending on the number of acceptor ions present.

Oxalic acid is a dibasic acid with two titratable protons, and in the complex anion each oxalate acts as a bidentate ligand occupying two **coordination sites** as illustrated.

By recognizing that the iron atom is in the Fe(III) oxidation state, we may simplify the proposed formula. Each oxalate anion carries two negative charges and the total charge on the complex anion is thus the sum of the charges on the total iron and oxalate ions present, or $2y - 3x$. The salt also contains only one complex anion associated with $wK^+$ cations, and as the whole complex is electrically neutral, it follows that $w = 2y - 3x$.

In this experiment you prepare, isolate, and recrystallize the complex salt. In a later experiment you will use permanganate titrations to find the simplest ratio $x:y$ and the empirical formula of the complex.

## PROCEDURE

### SAFETY NOTES

Acetone and ethanol are both flammable solvents. Take care not to expose their vapors to open flames.

Avoid handling beakers containing hot solutions with crucible tongs. Remember that such beakers may be safely handled by lifting at the rim within a few seconds of removing the heat source and may, if necessary, be supported in your hand on a towel.

Oxalic acid is a poisonous substance. Clean up all spills and place all residues in the marked disposal container.

### 1. Preparation of Ferrous Ammonium Sulfate Solution

Weigh 10 g of ferrous ammonium sulfate, $Fe(NH_4)_2(SO_4)_2 \cdot 6H_2O$, into a clean, dry 250-mL beaker on the rough balance. Add to the beaker 30 mL of warm distilled water to which 0.5 mL of 6$M$ sulfuric acid has been added.

### 2. Preparation of Iron(II) Oxalate

Measure 50 mL of the stock oxalic acid solution into a graduated cylinder and transfer it to the iron solution just prepared. A precipitate of yellow iron(II) oxalate, $FeC_2O_4$, will form. Add a boiling chip to the solution and carefully heat to boiling. The precipitate is very finely divided so guard against overheating and sputtering. Allow the solution to boil gently for a few minutes to help digest the precipitate—a process whereby the mean particle size is increased. Smaller particles dissolve more readily than larger ones because of their larger surface-to-volume ratio. Because the saturated solution is at equilibrium at the boiling temperature, as much material as dissolves must come out of solution to form solid. The bulk of solid formed deposits on pre-existing particles. The net effect is that small particles get smaller and large particles get larger. The larger the particle size, the more effectively is the solid freed from contaminants in the later wash.

Allow the solution to cool until the finely divided solid settles to the bottom of the beaker. Carefully decant away the supernatant liquid, retaining the precipitate in the beaker. Wash the precipitate in the beaker three times with 30-mL portions of hot distilled water, stirring, allowing to settle, and decanting off the supernatant in each case as before.

### 3. Oxidation to the Fe(III) State

Add 18 mL of $2M$ potassium oxalate solution to the ferrous oxalate precipitate in the beaker and heat the mixture to around 40°C. Prepare a buret with about 25 mL of 6% hydrogen peroxide stock solution. Slowly, and with constant stirring, add 17 mL of the peroxide to the oxalate preparation. Keep the temperature around 40°C, but be aware that the oxidation is an exothermic process and too rapid addition of peroxide will cause the mixture to overheat and lead to decomposition of the peroxide before it has a chance to react. Discard any remaining peroxide and rinse the buret thoroughly with distilled water.

### 4. Preparation of the Iron(III) Complex

Ready the buret with 15 mL of $1M$ oxalic acid. Heat the brown mixture from the preceding step to boiling, remove the burner, and quickly and with continuous stirring, add 9 mL of oxalic acid from the buret. Continue to stir the mixture and drop by drop add a further 4 mL of oxalic acid. The solution should now be a yellowish green color and if the work has been carefully done should be free from solid matter. If it is not, filter it free of solid through a glass filter funnel with a filter paper, discarding the residue and retaining the clear filtrate.

### 5. Crystallization of the Product

To the clear filtered solution from Step 4, contained in a 250-mL beaker, add 25 mL of ethanol. Redissolve any crystals that may form at this stage by gently heating the solution. Remember that ethanol vapor is flammable, so keep the heat gentle and have a 1-L beaker handy to quench any accidental fire. Cover the beaker containing the solution with aluminum foil and store it in your locker till the next laboratory period. Green crystals of the complex salt will deposit overnight. If no crystals form, try scratching the sides of the beaker with a glass stirring rod to promote nucleation.

At the start of the next period, set up a Buchner funnel and flask supported in a stand and clamp, and filter the crystalline product. Wash the crystals with two 10-mL portions of a 1:1 ethanol-water mixture by volume, and then twice with 10-mL portions of acetone. Continue aspiration until the crystals are dry and free from the odor of acetone. Remove the crystals and weigh them to the nearest 0.1 g on the rough balance. Note the result, and store the crystals in a 30-mL beaker completely wrapped in aluminum foil to keep exposure to light to a minimum. When you have determined the empirical formula of the complex, use this mass and the mass of ferrous ammonium sulfate used to calculate the percent yield of the salt.

Name _____ Instructor _____

# PRELABORATORY PREPARATION SHEET

1. What is meant by the following terms?

   a. Coordination complex _____

   _____

   b. Ligand _____

   _____

   c. Chelating agent _____

   _____

   d. Binuclear complex _____

   _____

   e. Anionic complex _____

   _____

2. Write a balanced net ionic equation for the preparation of iron(II) oxalate as described in Step 2 of the procedure.

   _____

3. The brown color of the precipitate formed on addition of $H_2O_2$ to the Fe(II) solution arises from the formation of ferric hydroxide, $Fe(OH)_3$. Why does this material form at this stage?

   _____

   _____

4. Why do you suppose ethanol is added to the solution of the complex in Step 5 of the procedure?

   _____

   _____

381

Name _____  Instructor _____

## DATA SHEET

### Checklist of operations; mark off as completed.

1. Mass of $Fe(NH_4)_2(SO_4)_2 \cdot 6H_2O$ taken  _____

2. Fe(II) solution prepared  _____

3. 50 mL $1M$ oxalic acid added  _____

4. Precipitate digested  _____

5. Supernatant decanted  _____

6. Precipitate washed  _____

7. $2M$ $K_2C_2O_4$ added  _____

8. 6% $H_2O_2$ added  _____

9. Buret washed  _____

10. $1M$ oxalic acid added  _____

11. Solution of complex filtered  _____

12. Ethanol added  _____

13. Solution covered and stored  _____

### Second period

14. Crystals filtered  _____

15. Crystals washed with $H_2O$/ethanol  _____

16. Crystals washed with acetone  _____

17. Crystals dried  _____

18. Mass of product  _____

Retain this sheet until you have determined the empirical formula of the complex and then calculate the percent yield. Hand in the sheet with your report for Experiment 20. No other report is needed for this experiment.

Name _____ Instructor _____

## DATA SHEET

### Checklist of operations; mark off as completed.

1. Mass of $Fe(NH_4)_2(SO_4)_2 \cdot 6H_2O$ taken  _____
2. Fe(II) solution prepared  _____
3. 50 mL 1$M$ oxalic acid added  _____
4. Precipitate digested  _____
5. Supernatant decanted  _____
6. Precipitate washed  _____
7. 2$M$ $K_2C_2O_4$ added  _____
8. 6% $H_2O_2$ added  _____
9. Buret washed  _____
10. 1$M$ oxalic acid added  _____
11. Solution of complex filtered  _____
12. Ethanol added  _____
13. Solution covered and stored  _____

### Second period

14. Crystals filtered  _____
15. Crystals washed with $H_2O$/ethanol  _____
16. Crystals washed with acetone  _____
17. Crystals dried  _____
18. Mass of product  _____

Retain this sheet until you have determined the empirical formula of the complex and then calculate the percent yield. Hand in the sheet with your report for Experiment 20. No other report is needed for this experiment.

# EXPERIMENT 19

# Standardization of a Potassium Permanganate Solution

### AIM

To determine the concentration of a stock solution of potassium permanganate by titration against an acidic solution of the primary standard sodium oxalate.

### EQUIPMENT

From stockroom:   50-mL buret
From locker:   30-mL beaker
                Erlenmeyer flasks (200-, 250-, or 300-mL)
                500-mL glass bottle
Laboratory:   Bunsen burner
                Ring or tripod stand
                Wire gauze
                Buret stand and clamp

### REAGENTS

Stock solution of potassium permanganate
0.75$M$ sulfuric acid
Anhydrous sodium oxalate

## BACKGROUND

### Associated Readings

The following sections of *Chemical Principles* should be read in conjunction with Experiment 19.

**Chapter 16** Oxidation-Reduction
    16.2 Oxidation-Reduction Processes in Aqueous Solution

Permanganate anion, $MnO_4^-$, is one of the strongest oxidizing agents available for quantitative use in the laboratory. In Experiment 17, where permanganate was used to reoxidize reduced vanadium species and to reconvert ferrous iron to the ferric state, we used the relative volumes of permanganate needed to establish the oxidation states of the reduced species. In doing so it was not necessary to know the exact concentration of the permanganate since we worked with the ratio of volumes of the same solution. However, to exploit fully the versatility of permanganate as a quantitative reagent in volumetric analysis, the concentration of the solutions used should be known.

The powerful oxidative action of permanganate, demonstrated by its ability to convert almost all other species to their highest oxidation states, imposes certain practical difficulties both in determining the concentration of its solutions and in its use as a reagent. Thus, although potassium permanganate, the most commonly available form of permanganate, may be prepared to a reasonable degree of purity, a solution of exact and stable concentration cannot be assumed when an exactly known mass of potassium permanganate is dissolved to make a solution of exactly known volume.

There are two principal reasons why this is so. First, even if the greatest care is taken, almost all aqueous solutions become contaminated during their preparation by small amounts of dust and bacteria from the air. Permanganate will oxidize nearly all organic compounds eventually to carbon dioxide and water, and until these oxidative processes have gone to completion in the permanganate solution its effective concentration will slowly and steadily diminish. Second, and more serious, permanganate will oxidize water in solutions which are even weakly acidic, according to the reaction:

$$4MnO_4^-(aq) + 2H_2O(aq) \rightarrow 3O_2(g) + 4MnO_2(s) + 4OH^-(aq) \tag{1}$$

These two conditions dictate that a solution of permanganate to be used for quantitative volumetric analysis must be made up some days, or even weeks, in advance of the date it is to be used; that it must be maintained at neutral or slightly basic pH; and that it must be stored in a tightly sealed container until it is ready for use.

Just as care must be taken in the preparation and storage of a permanganate solution, so also must care be exercised in its use as an analytical reagent. Only in acid solution is permanganate cleanly reduced to manganous ion, as indicated:

$$MnO_4^-(aq) + 8H^+(aq) + 5e^- \rightarrow Mn^{2+}(aq) + 4H_2O \tag{2}$$

In neutral or basic solution, a serious competing reaction is reduction of the permanganate to manganese dioxide:

$$MnO_4^-(aq) + 2H_2O + 3e^- \rightarrow MnO_2(s) + 4OH^-(aq) \tag{3}$$

which will proceed to an unpredictable extent. All of the titrations you will carry out with permanganate must be done with acidic solutions. It might seem that this would lead to problems with reaction (1), but the oxidation of water is a very slow process compared to the titration reaction (2), and does not seriously interfere in the short time needed to carry out a titration.

The rules for permanganate preparation and storage may be summarized:

**a.** Prepare the solution well in advance.
**b.** Store it in airtight glass containers.
**c.** Keep the pH of the stock solution neutral or *slightly* basic.
**d.** Carry out titrations only in acidic solution.

The standardization of a stored stable solution of permanganate is conveniently carried out by titration against oxalic acid. Sodium oxalate may be prepared to a very high degree of purity and, in its anhydrous form, is suitable for use as a primary standard. The salt may be made anhydrous by heating it for several hours at 120°C. It only very slowly reabsorbs water and so is reasonably stable over the period required to make a standard solution. Your instructor will give you directions for drying the salt if it has not been previously dried for you.

In acidic solution, such as is required for permanganate titrations, the oxalate anion–oxalic acid equilibrium greatly favors the undissociated parent weak acid, and oxidation of oxalic acid by permanganate proceeds according to the equation:

$$2MnO_4^-(aq) + 5H_2C_2O_4(aq) + 6H^+(aq) \rightarrow 2Mn^{2+}(aq) + 10CO_2(g) + 8H_2O$$

Hydrated protons are consumed in this reaction in large numbers, being converted to water, and it is important to maintain the acidity of the solution during the titration. Because the reaction is slow at room temperature, the titrations should be carried out at temperatures above 60°C.

## PROCEDURE

### SAFETY NOTES

Exercise care in handling hot solutions in beakers and flasks. Remember that a beaker or flask may normally be lifted safely at the rim within a half minute of removal of the heat source. Always test the rim temperature with your fingertips before trying to lift glassware in this way.

Oxalic acid and its salts are poisonous. Clean up all spills, disposing of waste solids in the labeled containers and flushing waste solutions down the sink with copious amounts of water.

Carry out the following steps for each of two oxalate samples, and for a third if your calculations indicate that the required degree of consistency in permanganate molarity has not been established.

1. Using the rough balance weigh about 0.25 g of dry sodium oxalate into a clean, dry 30-mL beaker. Transfer the beaker and its contents to the fine balance and weigh the assembly to the nearest mg. Transfer the sodium oxalate to a clean 250-mL Erlenmeyer flask, and reweigh the empty beaker to the nearest milligram to find the mass of oxalate transferred. Dissolve the oxalate in the flask by adding not less than 75 mL of 0.75$M$ sulfuric acid stock solution.
2. Ready your buret with potassium permanganate solution from your stoppered glass bottle. Pay particular attention to the proper procedures for rinsing the buret and for avoiding air bubbles in the tip.
3. Heat the flask containing the oxalate solution to about 85°C and titrate the oxalate solution against permanganate, taking care to keep the temperature above 70°C. If you should have to remove the thermometer from the flask during the titration, be sure to rinse any droplets of solution adhering to the bulb back into the flask. Record the

initial and final buret readings, estimating to the nearest 0.03 mL. Because there is no visible meniscus with permanganate, you should read the level corresponding to the flat upper surface of the reagent in the buret. Be sure to keep the buret in a strictly vertical position during the titration.

Complete the report sheet for this experiment before leaving the laboratory to be sure that you have a consistent set of results for the molarity of the permanganate. Your two estimates should agree to within $0.0008M$.

Name _____  Instructor _____

## PRELABORATORY PREPARATION SHEET

1. Why can we not prepare a standard solution of permanganate by simply dissolving a known mass of permanganate in a known volume of solution?

   _____
   _____
   _____
   _____

2. What precautions must be taken in storing a stock permanganate solution?

   _____
   _____
   _____

3. Why are permanganate titrations carried out in acid solution?

   _____
   _____
   _____
   _____

4. Why are most permanganate titrations carried out using hot solutions of the other reagent?

   _____
   _____

5. Permanganate reacts with oxalic acid, but we use a solution of sodium oxalate in the standardization. What allows for this apparent contradiction?

   _____
   _____

Name _____     Instructor _____

## DATA SHEET

### Weighings

Mass of beaker + oxalate    _____    _____    _____

Mass of beaker alone        _____    _____    _____

Mass of oxalate taken       _____    _____    _____

### Titrations

Final burret reading        _____    _____    _____

Initial buret reading       _____    _____    _____

Volume $KMnO_4$ delivered   _____    _____    _____

Name _____    Instructor _____

## DATA SHEET

### Weighings

Mass of beaker + oxalate    _____    _____    _____

Mass of beaker alone        _____    _____    _____

Mass of oxalate taken       _____    _____    _____

### Titrations

Final burret reading        _____    _____    _____

Initial buret reading       _____    _____    _____

Volume $KMnO_4$ delivered   _____    _____    _____

Name _____  Instructor _____

## REPORT SHEET

1. Enter the mass of sodium oxalate in each sample.

   (1) _____  (2) _____  (3) _____

2. Calculate the number of moles of sodium oxalate in each sample. Show one calculation in full, report only the results of the other.

   _____

   _____

   (1) _____  (2) _____  (3) _____

3. Give the equation for the reaction of permanganate with oxalic acid.

   _____

4. Calculate the number of moles of permanganate in each titration volume.

   **Titration volume**      **Moles**

   (1) _____      _____

   (2) _____      _____

   (3) _____      _____

5. Calculate the molarity of the permanganate solution for each titration.

   (1) _____

   (2) _____

   (3) _____

6. Give the mean molarity of your permanganate solution based on the above results. Average only molarities agreeing within $0.0008M$.

   Mean molarity _____ ± _____

## WORKSHEET

Carry out all the calculations for Experiment 19 on this sheet, attach it to your report sheet, and submit the two together. Do your calculations directly on this sheet. Do not do them on scratch paper and then copy them on to this sheet. Make no erasures and delete incorrect entries with a single line so as to leave the original data legible. Use a pen.

# EXPERIMENT 20

# Determination of the Empirical Formula of an Oxalatoferrate Complex by Permanganate Titration

### AIM

To establish the empirical formula of the previously synthesized potassium oxalatoferrate complex by permanganate titration.

### EQUIPMENT

| | |
|---|---|
| From stockroom: | 50-mL buret |
| From locker: | Erlenmeyer flasks (200-, 250-, or 300-mL) |
| | 1-L beaker |
| | 30-mL beaker |
| | Glass filter funnel |
| | Thermometer |
| | Washbottle |
| Laboratory: | Buret stand and clamp |
| | Tripod or ring stand |
| | Burner and wire gauze |
| | Filter paper or glass wool |

## REAGENTS

Standard permanganate solution
Concentrated sulfuric acid, $H_2SO_4$
2M and 6M sulfuric acid
Oxalatoferrate complex prepared in Experiment 18
Zinc metal granules

## BACKGROUND

### Associated Readings

The following sections of *Chemical Principles* should be read in conjunction with Experiment 20.

**Chapter 16** Oxidation-Reduction
    **16.2** Oxidation-Reduction Processes in Aqueous Solution
**Chapter 18** Coordination Compounds
    **18.1** Structure of Coordination Compounds

The complex prepared in Experiment 18 from potassium oxalate, oxalic acid, and ferrous ammonium sulfate has the composition

$$K_w[Fe_x(C_2O_4)_y] \cdot zH_2O$$

By noting that the iron atom is in the Fe(III) state and that the oxalate residues act as bidentate ligands, we deduced that the number of singly positively charged potassium ions, $w$, must equal $2y - 3x$, where $x$ and $y$ are the numbers of iron atoms and of oxalate residues in the complex ion. To establish the formula of the salt we need only determine the numerical values of $x$ and $y$. The number of $K^+$ ions follows from the relation given, and the number of molecules of water of hydration may then be found from the difference between the experimentally determined formula weight and that calculated from $x$, $y$, and $w$.

Of the four constituent species in the salt, only oxalate is directly titratable by permanganate. The amount of oxalate present is therefore found by carrying out a titration of the type used in the standardization of permanganate, the difference being that it is now the concentration of the oxalate that is sought and that of the permanganate that is known.

The iron content of the complex is determined by reducing the Fe(III) to Fe(II) with zinc metal. The Fe(II) produced in this way is then reoxidized to Fe(III) by permanganate titration and its concentration thus determined. Before this reduction can be carried out, the iron must be stripped of its complexing ligands. This is done by decomposing the oxalate residues with concentrated sulfuric acid, which breaks them down into the volatile gases carbon monoxide and carbon dioxide, and water. The remaining species in solution are $Fe^{3+}$ and $SO_4^{2-}$ ions. The sulfate ion is inert to permanganate and does not interfere with the titration.

## PROCEDURE

### SAFETY NOTES
Concentrated sulfuric acid is used in this experiment. Remember to handle this powerful reagent with proper care. Transfer it from the reagent dispenser

to your apparatus only at the hood and do not remove containers with the acid from the hood. Use the bicarbonate or carbonate powder provided to neutralize any spill. Should you get any acid on your skin, immediately wash it off with copious amounts of cold running water. Pay particular attention to the steps in the procedure involving this corrosive reagent and follow them exactly.

Two other concentrations of sulfuric acid, 2M and 6M, are also used in the experiment. Take care to choose the correct concentration of acid, carefully checking the labels on the reagent bottles.

### 1. Analysis for Oxalate

Using the rough balance, weigh about 0.2 g of the complex into a clean, dry 30-mL beaker. Transfer the beaker and its contents to the fine balance and weigh the assembly to the nearest milligram. Carefully transfer the complex into a clean 250-mL Erlenmeyer flask and reweigh the empty beaker to the nearest milligram to find the mass of complex transferred.

Add about 65 mL of distilled water to the flask and carefully, and slowly, add 10 mL of 6M $H_2SO_4$, gently swirling the flask to promote mixing of the contents. Use your wash bottle to rinse any solid adhering to the sides of the flask into the body of the solution. Heat the solution, while stirring, to around 80°C to dissolve the complex and to establish a temperature at which the titration reaction will proceed reasonably quickly. Take care to wash any drops of solution that may adhere to the stirring rod back into the flask if the rod is removed for any reason.

The solution is now ready to be titrated with permanganate. Carry out the titration as rapidly as possible, taking care to keep the temperature above 75°C.

### 2. Analysis for Iron

**NOTE: Read this section through once again before beginning work on this step. Observe proper precautions with concentrated $H_2SO_4$.**

Following the procedure outlined in Step 1, weigh out to the nearest milligram about 0.7 g of the complex into an Erlenmeyer flask.

Take the flask to the hood and carefully add to it 10 mL of concentrated $H_2SO_4$. In the hood, and using a tripod stand and gauze, gently warm the flask and its contents to decompose the complex anion. Do not overheat. You may assume decomposition is complete when the solution stops effervescing. Be sure that no unreacted solid is left adhering to the inside walls of the flask. Again working in the hood, carefully transfer the contents of this flask to another clean conical flask containing 25 mL of distilled water. Use a glass filter funnel for the transfer and wash the residue left in the first flask into the second with distilled water from your washbottle.

**NOTE: Do not add distilled water to the bulk of the reaction mixture in the first flask. Add distilled water only to rinse the residue remaining in the first flask after the original transfer.**

Make up the volume of solution in the second flask to about 60 mL. Swirl the flask to mix the contents thoroughly, and heat the solution for a few minutes to help discharge the turbidity. Now add about 7 g of zinc metal granules to the flask, swirl it to promote mixing, and with the washbottle rinse down the inside walls to ensure that all the zinc is in the main body of the solution.

Using a 1-L beaker as a water bath, and with the flask secured in a stand and clamp, heat the flask in boiling water for about 20 minutes to effect reduction to Fe(II). Test for complete reduction by removing one drop of the solution on the end of a stirring rod and mixing it with one drop of potassium thiocyanate solution [KSCN(aq)]. Any residual $Fe^{3+}$ ions will react producing a red color indicative of the formation of $Fe(SCN)^{2+}$ (see Experiment 1). Continue heating until reduction is complete.

When reduction is complete, cool the flask and its contents by holding it under the cold tap for a few seconds. Decant the solution through a glass filter funnel and filter paper (or glass wool plug) into a clean Erlenmeyer flask. Rinse the original flask and the filter bed with three 25-mL aliquots of $2M$ $H_2SO_4$, transferring the washings to the second flask. All zinc particles must be removed from the solution in this step.

Check that all zinc particles have been removed and that the solution is entirely clear. Heat the flask and its contents to about 80°C and carry out the permanganate titration.

Name _____  Instructor _____

## PRELABORATORY PREPARATION SHEET

1. Write a balanced net ionic equation for the reaction between ferrous iron and permanganate in acidic aqueous solution.

   _____

2. Write a balanced net ionic equation for the diagnostic reaction between ferric iron and thiocyanate ion.

   _____

3. The decomposition of the oxalate residues by concentrated sulfuric acid is a fairly complex process. However, assuming the products of reaction to involve $CO_2$, $CO$, and $H_2O$, write a simplified equation for the process.

   _____

4. Write a balanced net ionic equation for the reaction of permanganate and oxalic acid.

   _____

5. Enter here and in item 5 of the data sheet the mean molarity of your standard permanganate solution.

   _____ ± _____ M

Name _____  Instructor _____

## DATA SHEET

1. Oxalate analysis:

    Mass of beaker + complex  _____

    Mass of beaker alone  _____

    Mass of complex transferred  _____

2. Titration:

    Final buret reading  _____

    Initial buret reading  _____

    Volume $MnO_4^-$ delivered  _____

3. Iron analysis:

    Mass of beaker + complex  _____

    Mass of break alone  _____

    Mass of complex transferred  _____

4. Titration:

    Final buret reading  _____

    Initial buret reading  _____

    Volume $MnO_4^-$ delivered  _____

5. Mean molarity of the standard permanganate solution (from prelaboratory sheet)

    _____ ± _____

Name _____  Instructor _____

## DATA SHEET

1. Oxalate analysis:

   Mass of beaker + complex        _____

   Mass of beaker alone            _____

   Mass of complex transferred     _____

2. Titration:

   Final buret reading             _____

   Initial buret reading           _____

   Volume $MnO_4^-$ delivered      _____

3. Iron analysis:

   Mass of beaker + complex        _____

   Mass of break alone             _____

   Mass of complex transferred     _____

4. Titration:

   Final buret reading             _____

   Initial buret reading           _____

   Volume $MnO_4^-$ delivered      _____

5. Mean molarity of the standard permanganate solution (from **prelaboratory sheet**)

   _____ ± _____

Name _____ Instructor _____

# REPORT SHEET

1. Calculate the moles of oxalate present in the sample.

   _____
   _____
   _____

2. Calculate the mass of oxalate in the sample and express it as a mass percent of the sample.

   _____

3. Calculate the moles of iron present in the sample.

   _____
   _____
   _____

4. Calculate the mass and mass percent of iron in the sample.

   _____

5. Convert the mass percent data to mole percent data and determine an integer ratio $(x:y)$ for the iron to oxalate content of the complex.

   _____
   _____
   _____

6. From the charge relations between $x$, $y$, and $w$, evaluate $w$.

7. From $w$, find the mass percent of potassium in the complex by using the relationship:

$$\frac{\text{mass percent K}}{w \text{ (molar mass K)}} = \frac{\text{mass percent Fe}}{x \text{ (molar mass Fe)}}$$

   _____
   _____

8. Determine the mass percent of potassium in the complex by using the relationship analogous to that in Item 7 for oxalate.

   _____

   _____

9. Give the mean mass percent K obtained from Items 8 and 9.

   _____

10. Calculate the mass percent and mole percent of water in the complex and hence $z$.

    _____

11. Enter the empirical formula of the complex.

    _____

## WORKSHEET

Carry out all the calculations for Experiment 20 on this sheet, attach it to your report sheet, and submit the two together. Do your calculations directly on this sheet. Do not do them on scratch paper and then copy them on to this sheet. Make no erasures and delete incorrect entries with a single line so as to leave the original data legible. Use a pen.

# EXPERIMENT 21

# Kinetics of the Oxidation of Iodide Ion in Acidic Solution by Hydrogen Peroxide

### AIM

To determine the dependence of reaction rate on reactant concentration and temperature and to derive a reaction mechanism.

### EQUIPMENT

From stockroom:  5-mL and 10-mL graduated cylinders
From locker: Four 100-mL beakers
250-mL beaker
25-mL graduated cylinder
Laboratory: 50-mL graduated cylinder, ice bath

### REAGENTS

$1.0M$ hydrogen peroxide
$0.1M$ potassium iodide
$0.01M$ sulfuric acid
$0.02M$ sodium thiosulfate
Distilled water containing starch indicator

## OTHER MATERIALS

Stop watch or watch with seconds display
Ice for cooling

## BACKGROUND

### Associated Readings

The following sections of *Chemical Principles* should be read in conjunction with Experiment 21.

**Chapter 17** Chemical Kinetics
        **17.1** Rates of Reaction
        **17.2** Reaction Rates and Concentration
        **17.3** Rate Laws and Reaction Mechanisms
        **17.4** Reaction Rate and Temperature

Different chemical reactions take place at widely different rates, the time taken to reach equilibrium ranging from a few picoseconds to millenia. The rate at which a given reaction will occur is unpredictable and depends on a variety of conditions including the concentrations and physical states of the reactants, the temperature, and the presence of extraneous substances that may act as catalysts or inhibitors to speed up or retard the rate.

This uncertainty as to how fast a reaction will proceed may be contrasted with the comparative certainty with which thermodynamics can predict how far the reaction will proceed before reaching equilibrium. We can predict the extent of a reaction once it is initiated because we are dealing with thermodynamic state functions; we are concerned only with differences in the free energies, enthalpies, and entropies of reactants and products, quantities which are determinate.

We cannot predict the rate of a reaction because it depends not on the initial and final states of the reactants and products, but on the particular path taken in passing from one state to the other. Although almost all chemical processes involve collisions between molecules, we cannot predict the particular collision mechanisms that will be involved in a given reaction.

To understand how a particular reaction takes place at the molecular level we must determine experimentally how the reaction rate is affected by such factors as concentration of reactants and products, and by temperature. Such factors are referred to as the **kinetics** of the reaction.

For the generalized chemical reaction whose balanced equation is

$$a\text{A} + b\text{B} + \ldots \rightleftharpoons x\text{X} + y\text{Y} + \ldots$$

we can write an expression for the equilibrium constant

$$K = \frac{[\text{X}]^x[\text{Y}]^y \ldots}{[\text{A}]^a[\text{B}]^b \ldots}$$

Because chemical reactions involve collisions between reactant molecules, we might think that a rate expression for the reaction expressing its dependence on reactant concentration would be easy to predict, and that it would have the form

$$\text{rate} = k[\text{A}]^a[\text{B}]^b \ldots$$

Such a dependence of rate on concentration, however, implies that $a$ molecules of A and $b$ molecules of B collide simultaneously. Where $a$ and $b$ are greater than one, such collisions are rare and their frequency (which is calculable) quite inadequate to account for observed rates. The rate law for a reaction is indeed of the general form postulated, but the exponents for the concentration terms show no necessary connection with the stoichiometric coefficients of the balanced equation. Instead, the expression is

$$\text{rate} = k[A]^p[B]^q \ldots$$

where $k$, the **rate constant,** is independent of concentration at a given temperature, and the exponents $p$ and $q$ can only be found by experiment by observing how the rate varies as a function of each concentration.

The **reaction order** is defined as the sum of the exponents involved in the rate expression. For the generalized reaction above, the order would be $(p + q + \ldots)$. Many, though not all, reactions in which a single reactant decomposes to products are first order, e.g., the decomposition of $Cl_2O_7$ to chlorine and oxygen at 500 K.

$$2Cl_2O_7(g) \rightarrow 2Cl_2(g) + 7O_2(g)$$
$$\text{rate} = k[Cl_2O_7]$$

In a reaction following first-order kinetics, doubling the concentration of reactant doubles the rate of reaction.

Decomposition of $NO_2$ is second order, however,

$$2NO_2(g) \rightarrow 2NO(g) + O_2(g)$$
$$\text{rate} = k[NO_2]^2$$

Doubling reactant concentration in a second-order reaction quadruples the reaction rate.

The reaction

$$2NO(g) + O_2(g) \rightarrow 2NO_2(g)$$

the reverse of the second-order decomposition above, obeys the rate law

$$\text{rate} = k[NO]^2[O_2]$$

so that it is second order in NO, first order in oxygen, and third order overall. For a fixed concentration of oxygen, doubling the concentration of NO would quadruple the rate; for a fixed NO concentration, doubling that of oxygen would double the rate.

The exponents in the rate law need not be integers. Thus, for the reaction

$$CHCl_3(g) + Cl_2(g) \rightarrow CCl_4(g) + HCl(g)$$

the rate law is

$$\text{rate} = k[CHCl_3][Cl_2]^{1/2}$$

We have emphasized that these rate laws can only be determined experimentally by observing how rate alters as a function of each concentration at a given temperature, yet chemical intuition tells us that there should be a connection between the rate and the frequency of molecular collisions. Why can we not predict what this should be? Whereas we cannot predict the law from the chemical equation alone, we can use the observed kinetics to shed light on the mechanism of the chemical reaction. What we find helps to explain the difficulties of prediction.

Few chemical reactions actually occur at the molecular level as collisions between the numbers of molecules of reactants indicated by the stoichiometric coefficients. Instead the reactions proceed in a series of steps that may be summed to yield the overall equation. Each of these simpler **elementary reactions** reflects actual molecular behavior; the molecules do collide with one another in the numbers indicated by the stoichiometric coefficients. A rate law can therefore be written for each elementary reaction. The **molecularity** of the reaction, the number of reactant molecules involved, is the same as the order of the reaction. Not surprisingly, most elementary reactions involve collisions between only a pair of molecules and are **bimolecular. Unimolecular** processes are also fairly common, but trimolecular or higher order processes are rare.

The elementary reactions involved in more complex processes do not take place at the same rate; some are very fast, others very slow. The slowest elementary reaction determines the overall reaction rate and is known as the **rate-determining step.**

A study of the overall reaction kinetics often lets us postulate a plausible sequence of elementary reactions and rates that can account for the observed rate. It is a much less rewarding task to try to predict an overall rate in advance of experiment.

As an example of how a plausible mechanism can be proposed that is consistent with the observed kinetics, consider the reaction of order 3/2 described earlier:

$$CHCl_3(g) + Cl_2(g) \rightarrow CCl_4(g) + HCl(g)$$

for which the rate law is

$$\text{rate} = k[CHCl_3][Cl_2]^{1/2}.$$

This is a bimolecular reaction as written, but its overall order is 3/2. We are therefore sure that the reaction does not depend solely on a step in which a molecule of chlorine collides with one of chloroform. That is not an elementary reaction in this process.

A mechanism involving elementary reactions that is consistent with the observed kinetics can be proposed. Suppose the first step to be the fast dissociation of chlorine molecules into atoms, a process promoted by ultraviolet light.

$$Cl_2(g) \rightleftharpoons 2Cl(g) \tag{1}$$

Suppose the second step to be

$$Cl(g) + CHCl_3(g) \rightarrow HCl(g) + \cdot CCl_3(g) \tag{2}$$

where $\cdot CCl_3$ is a free radical, and suppose it to be slow, allowing (1) to reach equilibrium. Next suppose that the third step is a fast recombination of the $\cdot CCl_3$ radicals and Cl atoms

$$Cl(g) + \cdot CCl_3(g) \rightarrow CCl_4(g) \tag{3}$$

Note that the $\cdot CCl_3$, a vital intermediate, does not appear in the overall equation which is obtained by summing those for the elementary reactions.

The overall rate is determined by the slowest step and the rate law for that step is written to reflect the bimolecular process as

$$\text{rate} = k'[CHCl_3][Cl]$$

The unknown concentration [Cl] is obtained from the equilibrium constant for the dissociation reaction

$$K = \frac{[Cl]^2}{[Cl_2]}$$

as $[Cl] = (K[Cl_2])^{1/2}$ giving the rate of the rate determining step as

$$\text{rate} = k'K^{1/2}[CHCl_3][Cl_2]^{1/2}$$
$$= k[CHCl_3][Cl_2]^{1/2}$$

consistent with that observed for the overall rate. Note that simply obtaining a consistent rate expression is no guarantee that the proposed mechanism is indeed correct. Other confirmatory tests are usually needed.

In this experiment we study the effects of reactant concentration on the oxidation of iodide ion to tri - iodide ion by hydrogen peroxide in acidic aqueous solution and so determine the form of the rate law and the order of the reaction. From this data we postulate a mechanism for the reaction. We also study qualitatively the effect of change of temperature on the reaction rate.

The reaction proceeds according to the equation

$$H_2O_2(aq) + 3I^-(aq) + 2H^+(aq) \rightarrow I_3^-(aq) + 2H_2O$$

# Kinetics of the Oxidation of Iodide Ion in Acidic Solution by Hydrogen Peroxide

The rate equation for this reaction has the form

$$\text{rate} = k[H_2O_2]^p[I^-]^q[H^+]^r$$

where the exponents $p$, $q$, and $r$ are to be found by experiment. The exponents in this case are all integral so that we evaluate them by determining the rate of reaction as each reactant in turn is varied in concentration while the others are held fixed. It should be sufficient to determine the rate for three separate concentrations of each reactant.

The rate is followed by measuring the time needed to produce a fixed amount of $I_3^-$. We do this by adding to the reaction mixture a small fixed amount of thiosulfate ion that rapidly reacts with the tri-iodide ion to reconvert it to $I^-$ according to the equation

$$I_3^-(aq) + 2S_2O_3^{2-}(aq) \rightarrow 3I^-(aq) + S_4O_6^{2-}(aq)$$

Not until all of the thiosulfate is used up does a measurable amount of $I_3^-$ remain, and the presence of this ion is detected by using a starch indicator with which it forms an intense blue complex. The rate of reaction is therefore inversely proportional to the time taken for the blue color to appear in each reaction mixture.

## PROCEDURE

### SAFETY NOTES
There are no obvious chemical hazards associated with this experiment. Dispose of waste solutions by flushing them into the drains with cold water.

Students work in pairs, one mixing the reagents and the other timing the reaction. Reaction times should be measured to the nearest second.

A single stock solution of each of the three reagents is provided as well as a $0.02M$ solution of thiosulfate. The concentration of reactants is varied by using different volumes of each in a fixed volume of reaction mixture. The volume of the system is kept constant for all runs by addition of enough distilled water to maintain the chosen level. Because reaction rate is sensitive to temperature, the distilled water used is provided in special carboys so as to have equilibrated to room temperature. To simplify the bookkeeping, the starch indicator has been added to the distilled water.

You should experiment with the concentrations needed so that the time for a fixed volume of thiosulfate to be consumed may be accurately measured. As a starting point, use 5 mL of each reactant and 5 mL of thiosulfate made up to a total volume of 50 mL with distilled water. Be sure to use the same amount of thiosulfate in each subsequent run. If the reaction conducted under these conditions proceeds in under 200 seconds, then either use a larger volume of thiosulfate or increase the total reaction volume.

To study the effect of reactant concentration you should determine the rate for at least three different concentrations of each reactant in turn, with the remaining concentrations held constant. You should also conduct experiments in which more than one reactant concentration is varied at a time, but only to check your initial conclusions. To test the effect of temperature on rate, conduct two more experiments for which you already have room temperature data; one at around 0°C by cooling all the reactants and distilled water beforehand in an ice-water mixture, and the other at around 40°C by warming the ingredients to that temperature before the addition of room-temperature hydrogen peroxide.

### Setting up a Run

Collect about 50 mL of thiosulfate, 100 mL each of iodide, sulfuric acid and peroxide in separate clean, dry labeled beakers. That should be enough for all the runs.

Assume a reaction volume of 50 mL has been chosen. Measure out the required volume of iodide, sulfuric acid, and thiosulfate into a clean, dry 100-mL beaker or 50-

mL graduated cylinder. Use the smaller graduated cylinders and make the measurements of volume as accurately as possible. Note the sum of these volumes and mentally add to it the volume of peroxide which you intend to use. Do not add the peroxide to the reaction mixture at this stage. Now subtract the total volume of reactants, thiosulfate, and peroxide from the chosen total reaction volume. The difference is the amount of distilled water that you should now add to the reactants. Enter the volume of each reactant on your data sheet and check your arithmetic before going on.

Now ready the desired volume of $1M$ hydrogen peroxide in a separate graduated cylinder reserved for this reagent and, by mutual arrangement, with stirring add the peroxide rapidly to the reaction vessel and start the timer. The timer is stopped as soon as the blue color of the starch-iodine complex appears. The color appears quite suddenly so attend the process very carefully. Times should be recorded to the nearest second.

Name _____  Instructor _____

## PRELABORATORY PREPARATION SHEET

1. To determine the order of the iodide-peroxide reaction we do not need to know the exact concentrations of the reactants used. To what aspect of the experimental procedure is this attributable?

   _____

   _____

   _____

2. Consider the following reaction taking place in solution

   $$A + B \rightarrow AB$$

   At 350 K the reaction is complete in 15 seconds, but at 280 K needs 150 seconds to reach completion with the same initial concentration of reactants. What is the ratio of the rates at the two temperatures: rate(280):rate(350)?

   _____

3. For the reaction in Item 2, when the concentrations of A and B are each $1M$, the rate of formation of AB is 0.10 mol/s at a given temperature. If the reaction is first order in A and second order in B, what is the rate constant $k$?

   _____

4. Under the conditions in Item 3, the concentration of B is raised to $2.5M$. What is the new rate of reaction?

   _____

5. Why do you suppose the hydrogen peroxide in the 40°C run is not heated to that temperature beforehand?

   _____

423

Name _____  Instructor _____

Partner _____

## DATA SHEET

| Run | Volumes of stock solutions used | | | | | | | |
|---|---|---|---|---|---|---|---|---|
| | $H_2O_2$ (1$M$) | KI (0.01$M$) | $H_2SO_4$ (0.01$M$) | $Na_2S_2O_3$ (0.02$M$) | $H_2O$ | Total volume | Time | Temp. |
| 1 | _____ | _____ | _____ | _____ | _____ | _____ | _____ | _____ |
| 2 | _____ | _____ | _____ | _____ | _____ | _____ | _____ | _____ |
| 3 | _____ | _____ | _____ | _____ | _____ | _____ | _____ | _____ |
| 4 | _____ | _____ | _____ | _____ | _____ | _____ | _____ | _____ |
| 5 | _____ | _____ | _____ | _____ | _____ | _____ | _____ | _____ |
| 6 | _____ | _____ | _____ | _____ | _____ | _____ | _____ | _____ |
| 7 | _____ | _____ | _____ | _____ | _____ | _____ | _____ | _____ |
| 8 | _____ | _____ | _____ | _____ | _____ | _____ | _____ | _____ |
| 9 | _____ | _____ | _____ | _____ | _____ | _____ | _____ | _____ |
| 10 | _____ | _____ | _____ | _____ | _____ | _____ | _____ | _____ |
| 11 | _____ | _____ | _____ | _____ | _____ | _____ | _____ | _____ |
| 12 | _____ | _____ | _____ | _____ | _____ | _____ | _____ | _____ |
| 13 | _____ | _____ | _____ | _____ | _____ | _____ | _____ | _____ |
| 14 | _____ | _____ | _____ | _____ | _____ | _____ | _____ | _____ |
| 15 | _____ | _____ | _____ | _____ | _____ | _____ | _____ | _____ |

Name _____  Instructor _____

Partner _____

## DATA SHEET

| Run | Volumes of stock solutions used | | | | | | | |
|---|---|---|---|---|---|---|---|---|
| | $H_2O_2$ (1$M$) | KI (0.01$M$) | $H_2SO_4$ (0.01$M$) | $Na_2S_2O_3$ (0.02$M$) | $H_2O$ | Total volume | Time | Temp. |
| 1 | | | | | | | | |
| 2 | | | | | | | | |
| 3 | | | | | | | | |
| 4 | | | | | | | | |
| 5 | | | | | | | | |
| 6 | | | | | | | | |
| 7 | | | | | | | | |
| 8 | | | | | | | | |
| 9 | | | | | | | | |
| 10 | | | | | | | | |
| 11 | | | | | | | | |
| 12 | | | | | | | | |
| 13 | | | | | | | | |
| 14 | | | | | | | | |
| 15 | | | | | | | | |

Name _____  Instructor _____

Partner _____

# REPORT SHEET

1. Effect of varying $[H_2O_2]$:

   | Run | $[H_2O_2]$ | Time |
   |-----|-----------|------|
   | ___ | ___ | ___ |
   | ___ | ___ | ___ |
   | ___ | ___ | ___ |

   $(t_3:t_2:t_1)$ as an integer ratio: _____

   Order of $[H_2O_2]$: _____

2. Effect of varying $[I^-]$:

   | Run | $[I^-]$ | Time |
   |-----|--------|------|
   | ___ | ___ | ___ |
   | ___ | ___ | ___ |
   | ___ | ___ | ___ |

   $(t_3:t_2:t_1)$ as an integer ratio: _____

   Order of $[I^-]$: _____

3. Effect of varying $[H^+]$:

   | Run | $[H^+]$ | Time |
   |-----|--------|------|
   | ___ | ___ | ___ |
   | ___ | ___ | ___ |
   | ___ | ___ | ___ |

   $(t_3:t_2:t_1)$ as an integer ratio: _____

   Order of $[H^+]$: _____

4. Enter any additional concentration results using the same format.

5. Effect of temperature on rate (copy results from Data Sheet):

   | Run | $[H_2O_2]$ | $[I^-]$ | $[H^+]$ | Time | Temp. | |
   |-----|-----------|--------|--------|------|-------|---|
   | ___ | ___ | ___ | ___ | ___ | ___ | ca 0°C |
   | ___ | ___ | ___ | ___ | ___ | ___ | ca 20°C |
   | ___ | ___ | ___ | ___ | ___ | ___ | ca 40°C |

   Ratio of rates (40°:20°:0°C) _____

**6.** Postulated mechanism:

Overall rate expression: _____
Postulate a mechanism for the reaction consistent with the overall rate expression you have found. Identify the elementary reactions and clearly indicate the rate-determining step in your scheme.

## WORKSHEET

Carry out all the calculations for Experiment 21 on this sheet, attach it to your report sheet, and submit the two together. Do your calculations directly on this sheet. Do not do them on scratch paper and then copy them on to this sheet. Make no erasures and delete incorrect entries with a single line so as to leave the original data legible. Use a pen.

# EXPERIMENT 22

# Synthesis of Some Polymers

### AIM

To synthesize some organic polymers by condensation and addition polymerization reactions and to study their properties.

### EQUIPMENT

From stockroom: Large watchglass
From locker: 50-, 250-, and 1000-mL beakers
25-mL graduated cylinder
Filter funnel
Test tube
Stirring rod
Laboratory: Glass tubing
Ring or tripod stand
Wire gauze burner
Filter paper
pH paper
Paper towels
Plastic disposable gloves
Meter stick

## REAGENTS

| | |
|---|---|
| For thiokol synthesis: | 10% solution of sodium sulfide |
| | Powdered sulfur |
| | Ethylene dichloride |
| | 5% soap solution |
| | Concentrated aqueous ammonia |
| | 20% acetic acid |
| | Distilled water |
| For nylon synthesis: | 2% solution of sebacoyl chloride |
| | Hexamethylene diamine |
| For catechol-formaldehyde resin: | Catechol (1,2-dihydroxybenzene) |
| | 37% aqueous formaldehyde |
| | $1M$ sodium hydroxide |
| For plexiglass synthesis: | Methylmethacrylate |
| | Benzoyl peroxide catalyst |

## BACKGROUND

### Associated Readings

This experiment introduces some organic chemical reactions in addition to those covered in Chapter 21 of *Chemical Principles*. You should read all of that chapter so as to master the basic language and ideas of organic chemistry and so enhance your appreciation of this experiment.

Modern civilization rests upon a foundation of plastics. Produced originally as substitutes for natural materials, plastics now are prepared for almost every function and affect almost every aspect of our daily lives. Whereas some of the more common plastics are familiar by their household names such as nylon, Teflon, Dacron, Formica, Bakelite, plexiglass, and polyethylene, others are less well known but no less valuable. Polymers have been developed for functions as diverse as allowing natural breathing underwater, lightweight corrosion-free insulation of high thermal and environmental stability (typified by "Kaptan," used for more than 80 miles of wiring protection in the Concorde supersonic aircraft), and as protection against gunshots (afforded by Dupont's "Kevlar," which pound for pound has five times the strength of tensile steel).

Plastics are examples of the chemical family of molecules known as **polymers**, giant molecules built up by chemical combination of simple basic units known as **monomers.** The chemical reactions by which monomers are converted into polymers are known as **polymerization reactions.** Two principal types of polymerization are recognized: **addition polymerization,** in which the monomers combine directly in a head-to-tail fashion and no other products are formed besides the polymer (this method is used in this experiment in the synthesis of plexiglass); and **condensation polymerization** where one or more small molecule species are formed as byproducts of the main polymerization reaction. The other three preparations in this experiment illustrate this type of reaction.

Polymers formed in such reactions may be individual **long-chain molecules** which, depending on their chemical nature, may be **flexible** or **inflexible.** Nylon is an example of a fairly flexible long-chain molecule. Again, depending on the chemical nature of the monomeric units, the chains may be joined together laterally, **crosslinked,** to yield interlocking sheets or three-dimensional aggregates. Finally, the molecules may be aligned in the polymer to yield a strong, semicrystalline product, or be **unaligned,** yielding a predominantly amorphous product.

## Synthesis of Some Polymers

Besides the chemical properties of the monomers, the physical processes by which the polymer is produced are important, different techniques being used to produce threads or sheets from the same starting material, for example. The physical and chemical properties of a polymer are a reflection of its underlying molecular structure, and by controlling the chemical and physical processes involved in polymerization, plastics may be produced with almost any desired property. More chemists are engaged in the production of plastics in this country than in any other single branch of chemistry.

As examples of the relation between structure and function: a sticky polymeric glue would most likely be made up from flexible, unbranched, unaligned molecules; a synthetic rubber would use flexible, crosslinked, unaligned molecules; a comb or other molded product would use inflexible, unbranched, and unaligned molecules; whereas a plastic wrap would most likely use inflexible, unbranched, but aligned molecules. Various other combinations readily suggest themselves.

## SYNTHESES

### Thiokol, a Synthetic Rubber

Demand for synthetic rubber was stimulated during World War II by the Japanese occupation of the natural rubber plantations of Southeast Asia. Although today's technology and products are far superior to early efforts, the synthesis of thiokol illustrates the principles involved in engineering a molecule for a desired application.

The monomeric units from which thiokol is produced are ethylene dichloride, $(Cl.CH_2.CH_2.Cl)$, and polysulfide anions, $(S-S-S-S)^{2-}$. The reaction is a condensation polymerization with chloride ions being formed as a by-product.

$$n(Cl-CH_2-CH_2-Cl) + n(S-S-S-S)^= \longrightarrow \left( -CH_2-CH_2-\underset{\underset{S}{|}}{\overset{\overset{S}{|}}{S}}-S- \right)_n + 2nCl^-$$

Thiokol exhibits the properties of a polymer made up from flexible long-chain molecules interwoven with one another. A certain amount of crosslinking involving the sulfur atoms on different chains greatly reduces the elasticity which such a polymer would otherwise have. **Vulcanization,** a process by which synthetic rubbers are heated with sulfur, is used to deliberately reduce elasticity and improve hardness. This technique was discovered and exploited long before the chemical and structural principles involved were understood.

In a polymer with flexible, interwoven, unbranched molecules, stretching a sample simply involves straightening out the tangled molecules. Relaxing the applied tension allows the molecules to return to their tangled state, but because the molecules are free to move relative to one another to some extent, the final shape of the sample need not be the same as the original. Indeed, repeated stretching of the sample may lead to complete untangling of the molecular threads and a consequent loss of elasticity. By crosslinking the molecules at random sites, the ease of relative motion is reduced so that when tension is released the molecules spring back to their original positions and the sample retains its shape. Too high a degree of crosslinking will, of course, eliminate elasticity and control of the crosslinking is a critical factor in producing a useful product.

## PROCEDURE

### SAFETY NOTES
Ethylene dichloride is a toxic, volatile, highly flammable liquid. All manipulations involving this material are to be carried out in the hoods. Keep open flames away from the liquid.

Remember the rules for handling hot beakers. Use a towel under the base of the beaker when needed in carrying it from place to place. On no account use crucible tongs to carry beakers.

## 1. Preparation of Polysulfide Solution

Dissolve 5 g of sodium hydroxide pellets in 100 mL of 10% sodium sulfide solution contained in a 250-mL beaker. Place the beaker containing the solution on a tripod stand with wire gauze screen and bring the contents to a boil. Slowly and with vigorous stirring add about 5 g powdered sulfur to the boiling solution. The solution will turn a deep red color and some sulfur may remain unreacted. Take care not to allow the volume of solution to be too much reduced during this step; make it up if necessary to the original level.

Filter off the unreacted sulfur by passing the solution through a glass filter funnel and filter paper supported in a ring stand, collecting the filtrate in a 1-L beaker. Wash out the 250-mL beaker with a small amount of distilled water and pass the washings through the filter into the 1-L beaker. Wash the excess sulfur in the filter paper in the same way. Make up the level of solution in the 1-L beaker to 400 mL with distilled water. Some sulfur may precipitate from the solution as it cools but this will normally redissolve when the solution is reheated in the next step. Discard the sulfur in the filter paper in accordance with your instructor's directions.

## 2. The Polymerization Reaction

In the fume hood, heat the polysulfide solution to 75°C and add 20 mL of the 5% soap solution in 5-mL increments with stirring. The temperature may rise above 75°C as each lot is added; allow the solution to cool to 75°C between additions. The soap acts as an emulsifying agent during the subsequent precipitation step, helping the aqueous and organic phases to mix.

*Slowly,* and with vigorous stirring (*don't use your thermometer as a stirrer, it is fragile and expensive*), add 20 mL of ethylene dichloride to the hot solution taking care that the temperature remains below 75°C during the addition. When all of the ethylene dichloride has been added, continue to stir the mixture vigorously until the color changes from deep red to milky white. This should take about 10 minutes and the latter stages of the stirring may be done outside the hood. When the color of the reaction mixture is acceptable, leave the beaker aside to allow the precipitate to settle undisturbed. This normally takes about 40 minutes, but may extend to over an hour if the preparation has been carelessly carried out. At the end of this period a yellow-white layer of synthetic rubber will have settled out at the bottom of the beaker and the supernatant liquid will be clear and colorless.

## 3. Coagulation of the Product

Decant off as much of the supernatant solution as possible into the special disposal buckets provided. Do not decant off waste liquors into the drains as polymer formation may take place and lead to blockages.

At the fume hood make up a mixture of about 30 mL of concentrated ammonia solution in 370 mL of distilled water. Still at the hood, resuspend the polymer preparation by adding the aqueous ammonia to the beaker containing the rubber and stir thoroughly. Now with vigorous stirring, add just enough 20% acetic acid solution to the suspension to make it slightly acid. Test the acidity by touching a drop of the solution on the end of the stirring rod to a *small* strip of pH paper. About 80 to 100 mL of acetic acid will be needed. Acidification coagulates the rubber into a ball at the bottom of the beaker. Discard the supernatant solution into the sink (the solution contains only ammonium acetate, a harmless salt readily degraded) and remove the solid ball and wash it well with cold running water. The ball may be gently molded. Its color and consistency will depend upon the degree of crosslinking and exact sulfur content. Show the final sample to your assistant after estimating its volume and entering that information on your combined data and report sheet.

### Nylon, a Synthetic Fiber

Nylon is one of the oldest and best-known synthetic fibers. Produced in quantity during World War II when the demand for parachutes placed a strain on scarce reserves of silk, the material survived to become a symbol of post-war prosperity in the form of women's stockings. Nylon is a polymeric amide, a **polyamide,** formed by condensation of an alkane having terminal amino groups ($-NH_2$) to another alkane with terminal acid chloride functionalities (–COCl) to produce amide linkages (–NHCO–).

$$H_2N-(CH_2)_6-NH_2 + ClC-(CH_2)_8-CCl \longrightarrow$$
$$\underset{\text{hexamethylene diamine}}{} \quad \underset{\underset{O}{\|}}{} \quad \underset{\text{sebacoyl chloride}}{} \quad \underset{\underset{O}{\|}}{}$$

$$\left[ -NH-(CH_2)_6-NHC-(CH_2)_8-C- \atop \phantom{-NH-(CH_2)_6-NH}\underset{O}{\|}\phantom{(CH_2)_8-}\underset{O}{\|}\phantom{C-} \right]_x + 2xHCl$$

The diamine used in our experiment is hexamethylene diamine, 1,6-diaminohexane, and the acid chloride is sebacoyl chloride, derived from sebacic acid.

Because the polymerizing functional groups are terminal, the main product is a long-chain polymer free of covalent crosslinking. Its physical properties will depend primarily on the lengths of the intervening methylene moieties.

## PROCEDURE

### SAFETY NOTES
Both hexamethylene diamine and sebacoyl chloride are caustic materials, which burn skin. Be sure to wear the disposable plastic gloves provided when carrying out this experiment. Wash off any material spilled on your hands with copious cold running water and police up all spills on your bench top or elsewhere promptly. Do not handle the nylon fiber with your bare hands until it has been thoroughly washed. Keep open flames away from the sebacoyl chloride solution and avoid breathing the vapors.

### 1. Setting up the Two-phase System

Collect 25 mL of the 2% sebacoyl chloride solution in a clean dry 50 mL beaker. Set up a glass filter funnel in a stand so that its tip extends just above the surface of the chloride solution. *Very gently* pour about 7 mL of the aqueous hexamethylene diamine solution through the funnel so that it forms a layer above the chloride solution. A very thin film of nylon should form at the interface between the two solutions and prevent further interaction, but it is important that no other mixing of the two phases take place.

### 2. Formation of a Nylon Filament

Form a small hook at the end of a piece of glass rod or tubing, using the Bunsen burner. The hook should be as fine as possible. When the hook has cooled, carefully pierce the nylon film at its center. Twirl the hook so as to catch the nylon, then slowly and gently, with a vertical motion, raise the hook and the attached nylon filament until about 15 cm of thread has been pulled out. Now loop the thread around the barrel of a test tube or 30-mL beaker and wind out as long a continuous thread of nylon as possible from the system. A good preparation will yield several meters of thread. When you have wound out a sufficient length, cut the filament with scissors and slowly unwind it into a 1-L beaker filled with *cold* water. Keep a grip on the end of the filament as it enters the wash and arrange it so that the filament may be easily removed from the wash. Let the fiber sit for about 10 min. Now grasp the end of the thread and raise it out of the wash with one hand while gently rolling it between the finger and thumb of the other hand so as to collapse the hollow thread and remove any occluded reagent.

Repeat the washing and rolling process and spread the thread out to dry on a length of paper towel. You should aim for as fine a thread as possible. When the nylon is quite dry use a meter stick to measure its length and weigh it on the fine balance to the nearest milligram. Have your instructor inspect your sample.

### 3. Cleaning up

When you have drawn as much fiber from the reaction vessel as is needed, take a stirring rod and vigorously mix the two layers together. Any unspent material will then coagulate as a ball of polymer. The excess liquid should be placed in the marked disposal jars and the nylon ball should be well rinsed with cold water, squeezed dry, and discarded in the solid-waste disposal can. Rinse your gloves thoroughly with cold water before removing them and dispose of them in the same way.

**On no account place unspent reagents in the sinks or drains.**

### Catechol-Formaldehyde Resin

Catechol (I) and formaldehyde (II) react to form a polymeric resin in a condensation reaction in which the elements of water are eliminated as the polymer forms. The reaction takes place readily in the presence of base as a catalyst. The resulting polymer is a liquid resin and may be used as a glue. The polymer is spread on the surface of the materials to be glued and they are then heated to cure or harden the polymer. Resins which polymerize further to solids when heated in this way are known as thermosetting resins. Once hardened, such resins do not soften on reheating and they find wide use as glues in the manufacture of plywoods and other laminates.

## Procedure

### SAFETY NOTES
Catechol is an irritant chemical and the procedure described below should be carried out at the hood. Rough balances will be placed in the hoods for weighing catechol.

### 1. Polymerization Reaction

Fill a 1 L beaker about one quarter full of water and set it on a stand and wire gauze with burner to come to a boil.

On a large watchglass at the hood weigh out about 5 g of catechol. Add to the catechol on the glass 3 mL of 1 M sodium hydroxide and 5 mL of the 37% formaldehyde solution.

Adjust the burner under the large beaker so as to maintain a gentle boil. Set the watchglass and its contents atop the beaker and stir the reaction mixture as it is warmed by the steam from the bath for some 25 minutes. Remove the watchglass and allow it to cool. The polymer formed should be a very sticky liquid.

### 2. Testing the Adhesion

Test the adhesive properties of the resin by applying a thin film of it to a piece of filter paper or a small block of wood and pressing the glued material to some other material. Place the objects to be glued in an oven at 115°C for about ten minutes. Remove the sample and inspect the result. Continue heating for another ten minutes if the polymer has not cured. Have your assistant inspect your work.

### Polymethylmethacrylate (Plexiglass)

Methylmethacrylate monomers (I) polymerize in the presence of an appropriate catalyst to form a clear transparent solid. This is an example of addition polymerization, no other product being formed save the polymer. New chemical bonds are formed between the monomers by rearrangement of the electrons formerly in the double bonds of the methylmethacrylate and no atoms are eliminated. The catalyst used in this case is benzoyl peroxide (II) which dissociates on heating to form benzoyl *free radicals*, $C_6H_5CO_2\cdot$, the dot signifying the presence of a reactive non-bonded electron.

$$C_6H_5\text{-}C(=O)\text{-}O\text{-}O\text{-}C(=O)\text{-}C_6H_5 \longrightarrow 2\,C_6H_5CO_2\cdot \quad \text{(note the dot indicating the free electron)}$$

(II)

$$C_6H_5CO_2\cdot + H_2C=\underset{\underset{\text{(I)}}{}}{\overset{CH_3}{C}}\text{—COOCH}_3 \longrightarrow C_6H_5CO_2:CH_2-\overset{\cdot}{\underset{CH_3}{C}}\text{—COOCH}_3$$

$$C_6H_5CO_2:CH_2-\underset{\cdot}{\overset{CH_3}{C}}\text{—COOCH}_3 + H_2C=\overset{CH_3}{C}\text{—COOCH}_3 \longrightarrow C_6H_5CO_2:CH_2-\underset{COOCH_3}{\overset{CH_3}{C}}-CH_2-\underset{COOCH_3}{\overset{CH_3}{\overset{}{C}}}\cdot-CH_3$$

## PROCEDURE

### SAFETY NOTES
Benzoyl peroxide is an unstable substance and, if mishandled, may decompose with explosive force. It is quite harmless in the small amounts used in this experiment but, to be quite safe, your instructor will dispense the amount you need at the hood.

### 1. Polymerization Reaction

Fill a 1-L beaker one-third full of tap water and bring the water to a gentle boil. While the water is heating, place methylmethacrylate in a test tube to a depth of about 5 cm. Take the test tube to the hood where your instructor will dispense the small amount of benzoyl peroxide needed as a catalyst. Stir the contents of the tube with a glass stirring rod to thoroughly mix in the catalyst and stand the tube upright in the water bath. Leave the stirring rod in place to act as a "lollipop stick". Allow the test tube to remain in the bath for about 15 minutes, until the contents are quite viscous. Add enough cold water to the bath to lower the temperature to about 65°C and remove the heat. Let the test tube stand in the bath undisturbed until the temperature drops to about 30°C. Finish the cooling by removing the test tube and running cold tap water over its outside. The polymer should be hard.

### 2. Retrieval and Testing

Wrap the test tube in a towel and carefully break it with a hammer or file so as to release the plexiglass lollipop. Holding the sample by the glass stick, carefully remove all glass fragments from the polymer surface with a spatula and carefully shake the towel free of glass. Manually test the consistency of the polymer at room temperature and then, after it has been warmed in hot water, assess the degree of softening and note the results. Polymers of this type are known as **thermoplastics.**

Have your instructor inspect the result.

Name _____   Instructor _____

## PRELABORATORY PREPARATION SHEET

1. Identify 12 different polymeric substances in your immediate environment. In each case give the common name for the polymer and state its form and function.

|  | Polymer | Form | Function |
|---|---|---|---|
| (E.g., | Formica | 1.5 mm sheet | Veneer) |
| 1. | _____ | _____ | _____ |
| 2. | _____ | _____ | _____ |
| 3. | _____ | _____ | _____ |
| 4. | _____ | _____ | _____ |
| 5. | _____ | _____ | _____ |
| 6. | _____ | _____ | _____ |
| 7. | _____ | _____ | _____ |
| 8. | _____ | _____ | _____ |
| 9. | _____ | _____ | _____ |
| 10. | _____ | _____ | _____ |
| 11. | _____ | _____ | _____ |
| 12. | _____ | _____ | _____ |

2. Could a synthetic rubber be made from sodium polysulfide and ethyl chloride ($C_2H_5Cl$)? Give a reason for your answer.

_____

3. The nylon synthesized in this experiment is known as nylon 6-10. Examine the reaction scheme and give the formula for the repeating unit of nylon 8-8.

_____

4. What is the difference between a thermoplastic and a thermosetting polymer?

_____

5. In what way do addition and condensation polymerization differ?

_____
_____
_____
_____

6. There is no covalent crosslinking in nylon. Can you suggest a possible form of crosslinking that might be present in this molecule?

___

___

7. Use your knowledge of organic chemistry to propose a polymerization reaction other than those given here that would lead to a noncrosslinked long-chain molecule with minimal flexibility.

Name _____  Instructor _____

## DATA AND REPORT SHEET

Record the required data for each of your polymer samples, devising your own format. Describe their physical properties (color, hardness, strength, thermal behavior, etc.) and any readily determinable chemical properties. What natural material is most nearly akin to each of your samples? Describe the principal differences between these natural materials and your synthetics.

## DATA AND REPORT SHEET

Record the required data for each of your polymer samples, devising your own format. Describe their physical properties (color, hardness, strength, thermal behavior, etc.) and any readily determinable chemical properties. What natural material is most nearly akin to each of your samples? Describe the principal differences between these natural materials and your synthetics.

# Appendix

### Design, Use, and Care of Volumetric Glassware

Volumetric analysis depends on the precise manipulation of accurately known volumes of solution. The basic SI unit of volume, the cubic meter ($m^3$), is too large for practical purposes in volumetric work. Instead chemists use the cubic decimeter, $dm^3$, more commonly known as the liter and designated by the symbol L. Most items of volumetric glassware are designed for yet smaller volumes of solution and are usually marked in milliliters (mL) or tenths of a milliliter. For all practical purposes, 1 mL is equivalent to 1 $cm^3$.

### Effect of Temperature on Volumetric Glassware

The volume of a given mass of liquid varies with temperature, as does the volume of the container that holds it. Volumetric glassware is calibrated to a standard temperature of 20°C, the average room temperature in laboratories in North America.

Dilute aqueous solutions such as you use in this course have a coefficient of thermal expansion of about 0.025%/°C. This means that a temperature variation of only about 5°C will measurably affect the precision of normal volumetric measurements. A 25.00-mL sample of a liquid taken at 5°C would have a volume of 25.10 mL at 20°C. As the reproducibility of most volume measurements is to one drop (0.05 mL), you can see that solutions to be used in volumetric work should be allowed to equilibrate to room temperature before aliquots are taken. An aliquot is the name given to any precisely measured sample of liquid.

The coefficient of expansion of the glass vessels is much lower than that of the solutions used, being 0.003%/°C or less. For the experiments that we do we need not be much concerned with the effects on our glassware of small variations in laboratory temperature. However, we use only tepid solutions to clean the glassware and never expose volumetric ware to high temperature. Rapid cooling may distort the glass and lead to a permanent change in volume.

## Apparatus for the Precise Measurement of Volume

Volumetric glassware is of two types, that designed to contain and that designed to deliver a precisely measured volume of solution.

In the first category we use the volumetric flask, a round, flat-bottomed vessel with a long neck on which a line is etched. When filled to that line, the fiducial line, the flask *contains* the specified volume. Volumetric flasks are used in the preparation of standard solutions (Experiments 12, 15, and 19) and in the dilution of samples to known volume before taking aliquot portions with a pipet (Experiment 16).

In the second category are the volumetric pipet and buret, which when filled to the mark will *deliver* the indicated volume of liquid. The pipet is used to deliver an integral number of milliliters of solution and we use pipets of 5 and 25 mL capacity. The 50-mL buret we use is calibrated in intervals of 0.1 mL and is used in titration work in which an initially unknown volume of solution is to be delivered and precisely measured. Liquids contained in either a pipet or a buret show a curvature of the surface known as a **meniscus.** The bottom of the meniscus is commonly used as a reference point in reading the liquid level. The meniscus on a buret is more exactly read by using a buret card, an opaque card divided by a straight edge into black and white sections. The use of such a card is shown in Figure A.1. All readings should be made with the meniscus at eye level, so as to avoid errors due to parallax.

## Using the Volumetric Flask

The 250-mL volumetric flask is cleaned with a tepid detergent solution made up by adding 10 mL of the stock solution provided to 200 mL of tap water. If the water in your area is particularly hard you may need to use distilled water in the cleaning solution, depending on the type of detergent used. Pour about 25 mL of the cleaning solution into the flask, stopper it and swirl the solution vigorously so as to wet all the inside surfaces. Remove the stopper, which should be separately cleaned, and empty out the cleaning solution. Repeat this process at least twice more and then rinse the flask at least three times with distilled water, using the same technique. Don't fill the flask completely full of either cleaning solution or distilled water, as this simply wastes material and is no more effective

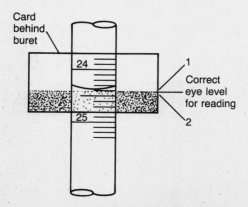

**Figure A.1**
Use of a buret card to read a buret.

than the technique described. At the end of the third rinse, invert the flask and allow it to drain. It should drain cleanly and there should be no beads of water on the inside surfaces. If beading occurs, repeat the wash-and-rinse cycle, using warm water if necessary for the wash.

To prepare a standard solution, add the exactly weighed mass of solid to the flask and add sufficient distilled water to dissolve the reagent. Normally the flask should not be more than half full at this stage. If the reagent was added to the flask through a glass filter funnel, make sure that all the solid is washed from the funnel into the flask by using your wash bottle with distilled water. Be sure to remove the funnel before going on. Stopper the flask and swirl its contents thoroughly to promote mixing. Make up the volume of solution to within a few millimeters of the fiducial line and then top up the level to the line by drop-by-drop addition of distilled water from the wash bottle.

It is very difficult to obtain a homogeneous solution in a volumetric flask. Because of the limited remaining air space in the flask at least 40 to 50 complete inversions of the stoppered flask are needed to ensure proper mixing. An alternative technique is to transfer the contents of the flask to your 500-mL glass bottle and to carry out the mixing in that. However, the glass bottle must be scrupulously clean and dry if used for this purpose, and if the solution is later returned to the volumetric flask, no attempt should be made to top it up to the mark. Some solution is inevitably lost in this process, and making up to the mark would only change the concentration of the solution.

When you have finished with the solution in the flask, dispose of it in accordance with the instructions given in the procedure for the experiment. Wash and rinse the flask in the same way as at the start of the experiment and return it to the stockroom. Don't wait until the last minute to clean up, and be sure to finish before the stockroom closes.

### Using the Volumetric Pipet

Liquid is drawn into the pipet by application of a slight vacuum created by a bulb. Never use your mouth to provide suction in pipetting, as you may accidentally ingest some solution.

The pipet is cleaned with the same tepid detergent solution described earlier. Draw in enough liquid to fill the bulb of the pipet (not the pipet bulb used for suction) about one-third full. Hold the pipet nearly horizontal and carefully rotate it so as to wet all the inside surfaces. Allow the pipet to drain and repeat the procedure at least twice more with fresh cleaning solution. Rinse the pipet with three separate lots of distilled water using the same technique. After the third rinse, the pipet should drain cleanly with no beads of water adhering to the inside surfaces. If water beads are present, repeat the wash-and-rinse cycle, using warm detergent solution if necessary, until the pipet drains cleanly. Wipe the outside dry with a paper tissue or a clean dry towel.

When the pipet is to be used to measure an aliquot of solution, it must first be conditioned with the solution in question. At the end of the cleaning and rinsing, the pipet was left with its inside surfaces wet with a thin film of distilled water and with a water droplet at its tip. These must be disposed of. Place a small quantity of the liquid to be measured in a clean, dry beaker. (To avoid contamination of your standard solution, the pipet should never be introduced directly into the volumetric flask.) Make sure that there is enough liquid to cover the stem of the pipet to a depth of at least 5 cm, and draw a small quantity of liquid into the pipet and use it to rinse the inside surfaces in the same way as in the cleaning cycle. Repeat this procedure at least twice more.

Now to another clean, dry beaker add at least one-and-a-half times as much solution as you intend to pipet and draw solution into the pipet some way beyond the fiducial mark. Quickly place your forefinger over the top of the pipet to stop the outflow of liquid. Check that there are no bubbles in the stem, tilt the pipet slightly from the vertical, and dry off the outside surface with a paper tissue. Touch the tip of the pipet to the edge of the glass beaker from which it was filled, and slowly allow the level of liquid to drop to the fiducial mark by gently relaxing the pressure on your forefinger. The bottom of

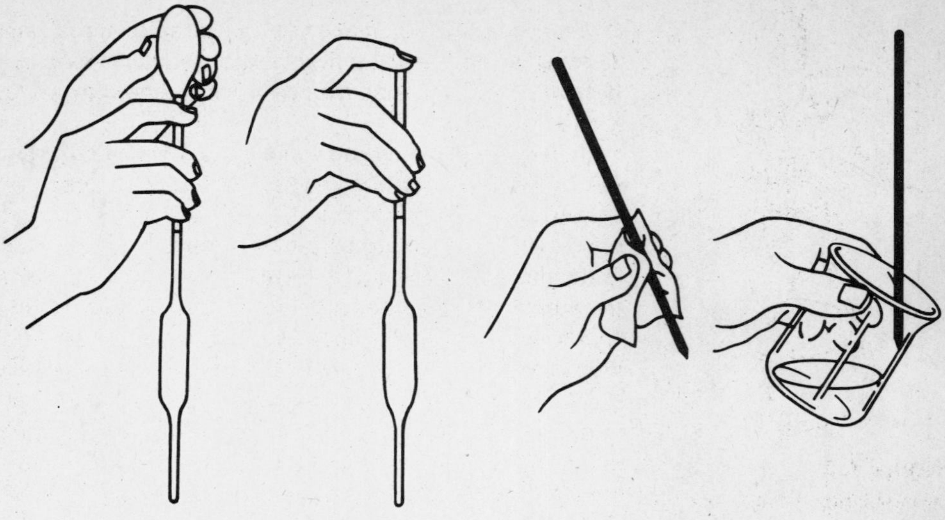

**Figure A.2**
Using the volumetric pipet.
**a** Draw liquid past the mark with a bulb
**b** Hold liquid above mark with forefinger
**c** Tilt pipet and wipe outside dry
**d** Adjust liquid level to the mark and allow the pipet to drain with tip in contact with inside wall.

the meniscus should be tangential to the mark when viewed at eye level. Place the tip of the pipet well inside the receiving vessel and allow the sample to drain. When free flow stops, rest the tip of the pipet against the wall of the vessel for a full 10 seconds. Finally withdraw the pipet with a gentle rotating motion to remove any droplet still adhering to the tip. Never attempt to blow out the small remaining volume of liquid inside the tip. Remember that the pipet is designed to deliver the stated volume, the volume it actually contains is larger by that final remaining volume. The steps in this process are shown in Figure A.2.

If a second aliquot of the same solution is to be measured, the pipet may be used without an intervening wash. However, if the pipet is to be used for another solution, the entire cleaning, rinsing, and conditioning process must be repeated and should be carried out in any event before the pipet is returned to the stockroom or to your locker. Be sure to return the pipet to the stockroom if it is not part of your permanent equipment.

## Using the 50-mL Buret

Like other volumetric ware, the buret must be scrupulously cleaned before being used. Clean the stem by adding about 10 mL of the prepared detergent solution to the buret with the stopcock closed. Repeatedly invert the buret so as to wet all the inside of the stem, and allow the detergent to drain out through the stopcock. Repeat this operation at least twice more and follow it with at least three distilled water rinses, using the same technique. Place the buret in a stand so that it is vertical and check that it drains cleanly. If water beads are present, repeat the wash-and-rinse cycle as necessary.

Most burets now have Teflon stopcocks that do not require lubrication but that do require cleaning. Remove the retaining ring, washer, spacer, and stopcock from the buret, wash and rinse them clean, dry them, and replace them in reverse order. Be particularly careful not to have any particles of dust or grit on the stopcock as they will cause a groove to form in the plastic as the stopcock is rotated. This will cause the buret to drip.

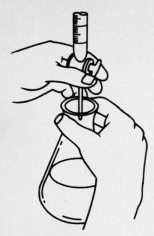

**Figure A.3**
Manipulating the stopcock on the buret.

If you have a dripping buret, first check the stopcock assembly to see that it is complete. Then tighten the retaining ring, but not to a point where it is difficult to turn the stopcock. If the buret continues to drip, have your instructor inspect it. If the stopcock is damaged it must be replaced.

If the stopcock requires lubricating (check with your instructor), use only the thinnest possible film of lubricant and be sure that none works its way into the stopcock hole or the tip.

To fill the buret, make sure that the stopcock is closed. Condition the buret with the solution to be used by adding about 10 mL of it to the buret and rotating and inverting the buret in the same way as for cleaning. Allow the liquid to drain through the tip. Repeat the procedure at least two more times. Now fill the buret to about 5 cm below the mark. If you used a glass funnel to transfer the solution into the buret, remove it at this time. Free the tip of air bubbles by rapidly rotating the stopcock and allowing small quantities of solution to pass. If air bubbles persist, remove the buret from the stand, place your thumb over the open end and invert the buret so that the tip is pointing up. Now open the stopcock and quickly reinvert the buret with the stopcock still open. The pressure of liquid falling should drive any remaining air bubble out of the tip. Reclose the stopcock and check that the volume of solution is still within about 5 mL of the mark. Read the exact level of solution in the buret. Don't waste time by trying to fill the buret exactly to the zero mark. All your data sheets have spaces for entries of initial and final readings. There is no merit in one of the readings being zero.

To manipulate the stopcock during a titration, use the technique illustrated by Figure A.3. That way the stopcock remains firmly seated. The stopcock should be tight enough to prevent leaks but not so tight as to be difficult to rotate.

In a titration, solution is added from the buret to a measured aliquot of reagent in an Erlenmeyer flask. The tip of the buret should be well inside the neck of the flask used for the titration. The solution should be introduced in increments of 1 mL or more and the flask should be swirled after each addition to promote mixing of the reactants. The size of the increments added is reduced as the end point is approached, until finally a drop-by-drop addition is used. When you think you are within a drop or two of the end point, rinse down the inside walls of the flask with distilled water from your wash bottle before completing the titration. Let 30 seconds pass between reaching the end point and reading the buret. This allows proper drainage of the buret stem and lets you check that the end point is indeed stable.

It is good technique when confronted by an unknown sample to carry out an initial rough titration in which you may overshoot the end point by bypassing the final dropwise addition. You can then add reagent to your subsequent titration samples much more rapidly, up to quite close to the end point, and so save a great deal of time.

Though a drop is normally the smallest increment added, smaller volumes can be added by allowing a small volume to form on the end of the tip and then touching the tip to the inside wall of the flask. That droplet may be combined with the bulk of the solution by tilting the flask and gently rotating it to absorb the droplet.

The buret is marked in 0.1-mL divisions. With care it may be read to 0.02 mL and certainly never to less than 0.05 mL.

The buret may be topped up as needed when the same solution is in use, but should always be cleaned, rinsed, and conditioned before use with a different solution and at the end of the laboratory period.

### Calibration of Volumetric Glassware

The design of volumetric glassware, with its allowance for reproducibility to the nearest drop or better, means that precise measurements are always possible. Accurate results can only be obtained, however, if the glassware actually contains (or delivers) the stated volume. To check the agreement between the actual and stated volumes, we calibrate the ware. Normally this involves obtaining an independent estimate of the volume by meas-

**TABLE A-1** Density of pure water (g/cm³) as a function of temperature (°C)

| Degrees | 0 | 1 | 2 | 3 | 4 | 5 | 6 | 7 | 8 | 9 |
|---|---|---|---|---|---|---|---|---|---|---|
| 15 | 0.999099 | 084 | 069 | 054 | 038 | 023 | 007 | *991 | *975 | *959 |
| 16 | 0.998943 | 926 | 910 | 893 | 877 | 860 | 843 | 826 | 809 | 792 |
| 17 | 774 | 757 | 739 | 722 | 704 | 686 | 668 | 650 | 632 | 613 |
| 18 | 595 | 576 | 558 | 539 | 520 | 501 | 482 | 463 | 444 | 424 |
| 19 | 405 | 385 | 365 | 345 | 325 | 305 | 285 | 265 | 244 | 224 |
| 20 | 203 | 183 | 162 | 141 | 120 | 099 | 078 | 056 | 035 | 013 |
| 21 | 0.997992 | 970 | 948 | 926 | 904 | 882 | 860 | 837 | 815 | 792 |
| 22 | 770 | 747 | 724 | 701 | 678 | 655 | 632 | 608 | 585 | 561 |
| 23 | 538 | 514 | 490 | 466 | 442 | 418 | 394 | 369 | 345 | 320 |
| 24 | 296 | 271 | 246 | 221 | 196 | 171 | 146 | 120 | 095 | 069 |
| 25 | 044 | 018 | *992 | *967 | *941 | *914 | *888 | *862 | *836 | *809 |
| 26 | 0.996783 | 756 | 729 | 703 | 676 | 649 | 621 | 594 | 567 | 540 |
| 27 | 512 | 485 | 457 | 429 | 401 | 373 | 345 | 317 | 289 | 261 |
| 28 | 232 | 204 | 175 | 147 | 118 | 089 | 060 | 031 | 002 | *973 |
| 29 | 0.995944 | 914 | 885 | 855 | 826 | 796 | 766 | 736 | 706 | 676 |
| 30 | 646 | 616 | 586 | 555 | 525 | 494 | 464 | 433 | 402 | 371 |

uring the mass of any given volume and deriving the actual volume from the known density of the sample at the prevailing temperature. Usually distilled water is used for this purpose. Because density varies with temperature, a table of the density of water as a function of temperature is needed. This is provided as Table A-1. In the simple calibration procedures that we use we will ignore such refinements as variation in the density of the air and calculate volume at a given temperature from mass measured at that temperature. A correction for thermal expansion is then used to calculate the volume of the ware at the standard temperature of 20°C.

### Example

A 25-mL pipet delivered a 24.90 g mass of water at 25°C. Calculate the volume delivered at 25°C and that delivered at 20°C.

The density of water at 25°C is 0.9970 g/mL, so that the volume of water at that temperature is 24.90 g/(0.9970 g/mL) or 24.97 mL. If the coefficient of expansion of the glass is 0.00003, then the volume delivered by the pipet at 20°C will be

$$V = 24.97 + (0.00003)(20 - 25)(24.97)$$
$$= 24.97 - 0.0037 \simeq 24.97 \text{ mL}$$

The outcome of the calculation confirms the minimal effect of variations in temperature on the volume of glassware, but does stress the larger effect of temperature on solution volume. Normally on a given day the laboratory temperature does not fluctuate much, and if the volumetric ware is accurate at 20°C, errors arising from volume changes due to temperature differences will tend to be self-canceling, as the different pieces of ware will respond similarly.

### Suggested Exercises

It is an excellent idea to practice the use of volumetric ware in advance of an actual experiment in which it is used to obtain accurate results. If time permits, as for example when you check into laboratory at the start of the second semester, you should carry out the following drills to familiarize yourself with the glassware and its use.

## 1. Prepare a Standard Solution of Sodium Chloride

Weigh a clean, dry 30-mL beaker to the nearest milligram. Take it to the rough balance and add to it about 5 g of sodium chloride. Reweigh the beaker and its contents to establish the exact mass of salt. Use the salt to prepare a standard solution in a 100- or 250-mL volumetric flask, following the directions given above.

## 2. Calibrate a 25-mL Pipet

Deliver the contents of a 25-mL pipet properly filled with distilled water into a previously weighed clean, dry 100-mL beaker and find the mass of water. Note the laboratory temperature and find the actual volume of water delivered, by using Table A-1. Then find the volume that would be delivered by the pipet at 20°C.

## 3. Titrate 0.1 *M* HCl Against 0.1 *M* NaOH

In separate clean, dry 250-mL beakers collect about 125 mL each of 0.1 $M$ HCl and 0.1 $M$ NaOH. Label the beakers so as to identify their contents. Prepare and set up a 50-mL buret and charge it with the NaOH. Into three clean Erlenmeyer flasks, pipet 25 mL aliquots of HCl. Titrate the samples. Use two drops of phenolphthalein as an indicator. The samples will turn purple when exactly enough NaOH has been added to neutralize the HCl present. You can get further information on titrations from Experiment 15, but need only note the appearance of a purple color stable for 30 seconds in these test runs. Make a first, rough titration and then see how closely the results of your second and third titrations agree with one another. With good technique they should agree to within 0.10 mL.

## WORKSHEET

Carry out all the calculations for these exercises on this sheet, attach it to your report sheet, and submit the two together. Do your calculations directly on this sheet. Do not do them on scratch paper and then copy them on to this sheet. Make no erasures and delete incorrect entries with a single line so as to leave the original data legible. Use a pen.

**Stockroom Slip**
**Experiment 1**
Buchner funnel and flask

Name _____

Date _____

**Stockroom Slip**
**Check-in Day**
Buchner funnel and flask

Name _____

Date _____

**Stockroom Slip**
**Experiment 4**
Buchner funnel and flask
Porcelain crucible and lid

Name _____

Date _____

**Stockroom Slip**
**Experiment 3**
Porcelain crucible and lid

Name _____

Date _____

**Stockroom Slip**
**Experiment 6**
Dumas apparatus (if supplied)

Name _____

Date _____

**Stockroom Slip**
**Experiment 5**
500-mL round-bottomed flask
Rubber tubing

Name _____

Date _____

**Stockroom Slip**
**Experiment 8**
5-mL pipet

Name _____

Date _____

**Stockroom Slip**
**Experiment 7**
2 styrofoam cups with lids
Narrow-range thermometer

Name _____

Date _____

**Stockroom Slip**
**Experiment 11**
Freezing point apparatus
Narrow-range thermometer

Name _____

Date _____

**Stockroom Slip**
**Experiment 9**
25-mL pipet
50-mL buret

Name _____

Date _____

**Stockroom Slip**
**Experiment 14(II)**
5-mL graduated cylinder

Name _____

Date _____

**Stockroom Slip**
**Experiment 12**
5-mL and 25-mL pipets
50 mL buret

Name _____

Date _____

**Stockroom Slip**
**Experiment 16**
5-mL and 25-mL pipets
50-mL buret
250-mL volumetric flask

Name _____

Date _____

---

**Stockroom Slip**
**Experiment 15**
25-mL pipet, 50-mL buret
250-mL volumetric flask

Name _____

Date _____

---

**Stockroom Slip**
**Experiment 18**
50-mL buret
Buchner funnel and flask

Name _____

Date _____

---

**Stockroom Slip**
**Experiment 17**
50-mL buret

Name _____

Date _____

---

**Stockroom Slip**
**Experiment 20**
50-mL buret

Name _____

Date _____

---

**Stockroom Slip**
**Experiment 19**
50-mL buret

Name _____

Date _____

---

**Stockroom Slip**
**Experiment 22**
Large watchglass

Name _____

Date _____

---

**Stockroom Slip**
**Experiment 21**
5-mL and 25-mL graduated cylinders
50-mL buret

Name _____

Date _____

---

**Stockroom Slip**

Name _____

Date _____

---

**Stockroom Slip**
**Volumetric Drill**
25-mL pipet, 50-mL buret
250-mL volumetric flask

Name _____

Date _____

---

**Stockroom Slip**

Name _____

Date _____

---

**Stockroom Slip**

Name _____

Date _____

## Table of Atomic Weights

| Element | Symbol | Atomic Number | Atomic Weight | Element | Symbol | Atomic Number | Atomic Weight |
|---|---|---|---|---|---|---|---|
| Actinium | Ac | 89 | 227.0278* | Neodymium | Nd | 60 | 144.24 |
| Aluminum | Al | 13 | 26.98154 | Neon | Ne | 10 | 20.179 |
| Americium | Am | 95 | (243) | Neptunium | Np | 93 | 237.0482* |
| Antimony | Sb | 51 | 121.75 | Nickel | Ni | 28 | 58.70 |
| Argon | Ar | 18 | 39.948 | Niobium | Nb | 41 | 92.9064 |
| Arsenic | As | 33 | 74.9216 | Nitrogen | N | 7 | 14.0067 |
| Astatine | At | 85 | (210) | Nobelium | No | 102 | (259) |
| Barium | Ba | 56 | 137.33 | Osmium | Os | 76 | 190.2 |
| Berkelium | Bk | 97 | (247) | Oxygen | O | 8 | 15.9994 |
| Beryllium | Be | 4 | 9.01218 | Palladium | Pd | 46 | 106.4 |
| Bismuth | Bi | 83 | 208.9804 | Phosphorus | P | 15 | 30.97376 |
| Boron | B | 5 | 10.81 | Platinum | Pt | 78 | 195.09 |
| Bromine | Br | 35 | 79.904 | Plutonium | Pu | 94 | (244) |
| Cadmium | Cd | 48 | 112.41 | Polonium | Po | 84 | (209) |
| Calcium | Ca | 20 | 40.08 | Potassium | K | 19 | 39.0983 |
| Californium | Cf | 98 | (251) | Praseodymium | Pr | 59 | 140.9077 |
| Carbon | C | 6 | 12.011 | Promethium | Pm | 61 | (145) |
| Cerium | Ce | 58 | 140.12 | Protactinium | Pa | 91 | 231.0359* |
| Cesium | Cs | 55 | 132.9054 | Radium | Ra | 88 | 226.0254* |
| Chlorine | Cl | 17 | 35.453 | Radon | Rn | 86 | (222) |
| Chromium | Cr | 24 | 51.996 | Rhenium | Re | 75 | 186.207 |
| Cobalt | Co | 27 | 58.9332 | Rhodium | Rh | 45 | 102.9055 |
| Copper | Cu | 29 | 63.546 | Rubidium | Rb | 37 | 85.4678 |
| Curium | Cm | 96 | (247) | Ruthenium | Ru | 44 | 101.07 |
| Dysprosium | Dy | 66 | 162.50 | Samarium | Sm | 62 | 150.4 |
| Einsteinium | Es | 99 | (252) | Scandium | Sc | 21 | 44.9559 |
| Erbium | Er | 68 | 167.26 | Selenium | Se | 34 | 78.96 |
| Europium | Eu | 63 | 151.96 | Silicon | Si | 14 | 28.0855 |
| Fermium | Fm | 100 | (257) | Silver | Ag | 47 | 107.868 |
| Fluorine | F | 9 | 18.998403 | Sodium | Na | 11 | 22.98977 |
| Francium | Fr | 87 | (223) | Strontium | Sr | 38 | 87.62 |
| Gadolinium | Gd | 64 | 157.25 | Sulfur | S | 16 | 32.06 |
| Gallium | Ga | 31 | 69.72 | Tantalum | Ta | 73 | 180.9479 |
| Germanium | Ge | 32 | 72.59 | Technetium | Tc | 43 | (98) |
| Gold | Au | 79 | 196.9665 | Tellurium | Te | 52 | 127.60 |
| Hafnium | Hf | 72 | 178.49 | Terbium | Tb | 65 | 158.9254 |
| Helium | He | 2 | 4.00260 | Thallium | Tl | 81 | 204.37 |
| Holmium | Ho | 67 | 164.9304 | Thorium | Th | 90 | 232.0381* |
| Hydrogen | H | 1 | 1.00794 | Thulium | Tm | 69 | 168.9342 |
| Indium | In | 49 | 114.82 | Tin | Sn | 50 | 118.69 |
| Iodine | I | 53 | 126.9045 | Titanium | Ti | 22 | 47.90 |
| Iridium | Ir | 77 | 192.22 | Tungsten | W | 74 | 183.85 |
| Iron | Fe | 26 | 55.847 | Unnilennium | Une | 109 | (266) |
| Krypton | Kr | 36 | 83.80 | Unnilhexium | Unh | 106 | (263) |
| Lanthanum | La | 57 | 138.9055 | Unnilpentium | Unp | 105 | (262) |
| Lawrencium | Lr | 103 | (260) | Unnilquadium | Unq | 104 | (261) |
| Lead | Pb | 82 | 207.2 | Unnilseptium | Uns | 107 | (262) |
| Lithium | Li | 3 | 6.941 | Uranium | U | 92 | 238.029 |
| Lutetium | Lu | 71 | 174.967 | Vanadium | V | 23 | 50.9415 |
| Magnesium | Mg | 12 | 24.305 | Xenon | Xe | 54 | 131.30 |
| Manganese | Mn | 25 | 54.9380 | Ytterbium | Yb | 70 | 173.04 |
| Mendelevium | Md | 101 | (258) | Yttrium | Y | 39 | 88.9059 |
| Mercury | Hg | 80 | 200.59 | Zinc | Zn | 30 | 65.38 |
| Molybdenum | Mo | 42 | 95.94 | Zirconium | Zr | 40 | 91.22 |

The atomic weights of many elements are not invariant. They depend on the origin and treatment of the material. The values given here are from a 1977 report of the Commission on Atomic Weights of IUPAC. These values apply to elements as they exist on earth and to certain artificial elements. Values in parentheses are used for radioactive elements whose atomic weights cannot be quoted precisely without knowledge of origin. The value given is the mass number of the isotope of longest known half-life.

* Relative atomic mass of the most common long-lived isotope.